COLONEL FREMONT AND THE AUTHOR TAKING ASTRONOMICAL OBSERVATIONS.—P. 129.

Incidents of Travel and Adventure in the Far West

with Colonel Fremont's Last Expedition

ACROSS THE ROCKY MOUNTAINS: INCLUDING THREE
MONTHS' RESIDENCE IN UTAH, AND A PERILOUS TRIP
ACROSS THE GREAT AMERICAN DESERT TO THE PACIFIC

Solomon Nunes Carvalho

Introduction by Ava F. Kahn

UNIVERSITY OF NEBRASKA PRESS
LINCOLN AND LONDON

Introduction and map © 2004 by the Board of Regents of the
University of Nebraska
All rights reserved
Manufactured in the United States of America

First Nebraska paperback printing: 2004

Library of Congress Cataloging-in-Publication Data
Carvalho, Solomon Nunes, 1815–1897.
Incidents of travel and adventure in the Far West: with Colonel
Fremont's last expedition across the Rocky Mountains, including three months' residence in Utah, and a perilous trip across
the great American desert to the Pacific / by Solomon Nunes
Carvalho; introduction by Ava F. Kahn.
p. cm.
Originally published: New York: Derby & Jackson, 1858.
ISBN 0-8032-6444-5 (pbk.: alk. paper)
1. West (U.S.)—Description and travel.   2. Frontier and
pioneer—West (U.S.)   3. West (U.S.)—History—1848–1860.
4. Carvalho, Solomon Nunes, 1815–1897.   5. Frâemont, John
Charles, 1813–1890.   6. Mormon Church.   I. Title.
F593.C283   2004
978'.02—dc22
2004001062

This Bison Books edition follows the original in beginning
chapter 1 on arabic page 17; no material has been omitted.

INTRODUCTION

Ava F. Kahn

> On the 22d August 1853, after a short interview with J. C. Frémont, I accepted his invitation to accompany him as artist of an Exploring Expedition across the Rocky Mountains.
> —Solomon Nunes Carvalho

Solomon Nunes Carvalho (1815–97) and John Charles Frémont (1813–90), men with very different ambitions, education, and heritage, spent six months together crossing the Great Plains and the mountainous West in the cold winter of 1853–54. Frémont, on his fifth expedition, wanted to establish once and for all the practicality of the central route for a continental railway; Carvalho, an artist, was there to document the vistas, not only by sketching but also by using the new medium of photography.

Both Carvalho and Frémont shared a South Carolina childhood. Both were independent, class conscious, and headstrong; they shared an interest in scientific inventions, experimentation, and travel. Most of all, both hated to fail. Carvalho's writing brings to life some of their successes and describes their shared hardships but leaves many questions about the purpose of the book, *Incidents of Travel and Adventure in the Far West with Colonel Fremont's Last Expedition*, and the survival of the expedition photographs.

Carvalho had many successes in later life, but the expedition and the subsequent publication of his book established his principal identity for future generations. Frémont published only short excerpts from his fifth expedition journal in his 1887 book, *Memoirs of My Life*, establishing

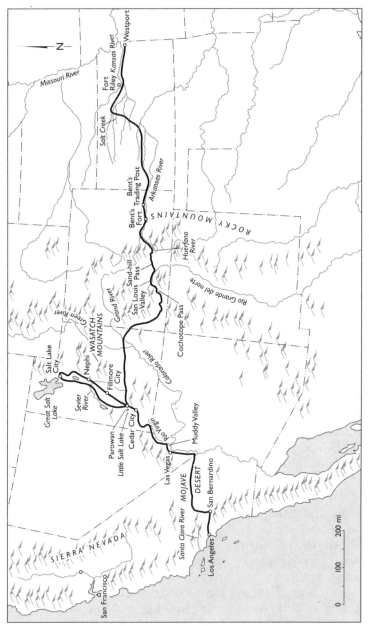

The route of Solomon Nunes Carvalho, 1853–54

Carvalho's 1856 book as the most important description of the journey. Although his words survived, his photographs did not. Ironically, Carvalho was remembered as "the Pathfinder's photographer."

Incidents of Travel can be divided into three sections: Carvalho's experiences with the expedition, his thoughts on Utah's Mormons, and a short description of his life in California. Written in the years between the California gold rush and the Civil War, his narrative gives insight into the daily lives of expedition members, the hardships and joys of field photography, the relations between the American Indians and exploration parties, the effect of the harsh western environment on the expedition, and the practices of the Mormon community that rescued him from the Rockies. Carvalho's perspective is unique, as he was the first Jewish writer to publish impressions of a western expedition, and his observations reflect his Jewish education and observance. Because he comments on such things as expedition members eating unkosher animals, Carvalho's writings were distinct from other western expedition narratives.

When Carvalho was born in Charleston, South Carolina, in 1815, the city was already home to a strong American–Sephardic Jewish community. Most of its six hundred Jews had come to Charleston from the West Indies and England. They came to Charleston because of the religious freedom and economic opportunity it offered. An important Jewish community since before the American Revolution, the city was a primary American port, where Jews were traders and merchants.

Carvalho's ancestors were Sephardic, with roots in Spain and Portugal (as a Spanish-speaking Jew, Carvalho, when necessary, served the expedition as a translator). His father, David Nunes Carvalho—born and educated in London—was a merchant and a Jewish educator who authored plays and hymns.[1] His mother, Sarah D'Azevedo, was also from a Sephardic family with roots in Amsterdam, London, and the West Indies.

Educated in Jewish tradition and learning and exposed to science, the classics, arts, and American patriotism, Carvalho acquired a strong Jewish and American identity. He was not only a comfortable but an active participant in both the secular and Jewish worlds. Known as an artist, inventor, photographer, businessman, and author, Carvalho also was a supporter of Jewish education, the founder of synagogues and benevolent societies, and a writer on Jewish subjects. As an American-born Jew, he was equally at ease discussing his inventions and photography with local experts or debating interpretations of Jewish practice with the scholars of the day. Carvalho was fully integrated into American life and in the belief that as a people, Americans were unstoppable. He also shared in the country's fascination with the unsettled West and with science.

Carvalho believed in the importance of education. It was vital to him to educate Jewish children in English through the use of English hymns and sermons, thus making Judaism accessible to all American-born Jews. For Carvalho, no separation existed between his religious and secular life. He believed that "[r]eligion must signify itself in our action in life, ay, it must embrace the whole sphere of our activities and affections."[2]

By the mid-1830s, Carvalho decided that his true calling was in the arts. Charleston was home to other Jewish artists, and although some felt that Biblical law restricted making images of people, Carvalho's work was accepted by the Jewish and non-Jewish communities.[3] His first activities for Charleston's Jewish community involved art; when the synagogue was destroyed by fire in 1838, Carvalho, who was in Philadelphia at the time, created a painting from memory of the synagogue's interior. He sent his painting to Charleston and asked to be compensated. Validating his artistic talent, they paid him fifty dollars.

In 1845 he married Sarah Miriam Solis, the daughter of a prominent Philadelphia Sephardic family with roots back to the American Revolution. Isaac Leeser (1806–68), publisher of the *Occident*, translator of the Hebrew Bible into

English, *hazzan* of Congregation Philadelphia's Mickveh Israel, and friend of both Carvalho and the Solis family, officiated at the ceremony. The couple made their first home in Philadelphia, where Carvalho listed himself as a portrait painter.

In the 1840s and 1850s, along with portrait painting, Carvalho experimented with Biblical subjects and landscape—and daguerreotype photography.[4] In fact, he may have been the first Jewish photographer in the United States. Carvalho was first attracted to the medium as a way to aid his work as a painter; he would use a photograph as a template for a later painting. Portrait photography eventually became part of his commercial offerings, marrying his interests in science and art. In advertisements for his portrait studios he stressed his "scientific" knowledge of the process of photography.[5] Between 1849 and 1853, he experimented with different photographic techniques and made modifications to his methods. One of his innovations was the application of a transparent enamel or varnish to protect his daguerreotypes, thus eliminating the need for cases or glass covers and making the daguerreotypes more durable, lighter, and easier to transport.

Carvalho's success and reputation as a daguerreotypist gained the attention of Frémont, who was in New York planning his fall expedition west.[6] Having himself tried and failed to produce daguerreotypes on two previous expeditions, Frémont decided to hire an experienced professional to join his scientific corps. Jessie Benton Frémont, the explorer's wife, later recalled, "In New York the daguerre apparatus was bought, and a good artist secured, Mr. Carvalho."[7]

Although a professional and an innovator, Carvalho's experience was in the studio, not in the outdoors and certainly not in the frozen wilderness up to his waist in snow. In the early 1850s daguerreotype was at its height in popularity, and it had been successfully accomplished in the field but never on a winter expedition to the West.[8] Given ten days to purchase equipment and leave New York to meet

Frémont in St. Louis, Carvalho consulted an expert, Edward Anthony, who was a pioneer in outdoor photography.[9] Having worked with a government survey to document the northeastern boundary of the United States, Anthony was familiar with photography in primitive conditions. Anthony's photographic supply house provided Carvalho with the chemicals and specially designed equipment that he would need. Other photographers thought Carvalho would fail, as daguerreotypes had never been made in the mountains in the winter, but Carvalho later stated that he "found no such word" in his vocabulary.[10] With that bold mind-set, he set off to join Frémont.

Frémont had led four previous expeditions—all scientific surveys to the West—in 1842, 1843, 1845, and 1848. His third expedition had resulted in his court-martial for mutiny and other charges arising from a clash with a superior officer in the wake of the U.S. conquest of California during the Mexican-American War. His fourth expedition had ended in tragedy when ten men died of the effects of cold and starvation during their winter trek though the Rockies. Several of the expedition's survivors as well as many in Washington questioned Frémont's leadership abilities.[11] By the fifth expedition, Frémont was a controversial figure.

Frémont privately funded the fifth expedition, as Jefferson Davis, secretary of war and a supporter of the southern route, did not select him to lead one of three government-supported railway survey expeditions. Like the fourth, Frémont's fifth expedition sought to find a practical route along the thirty-eighth parallel for the transcontinental railroad. Frémont needed to prove that the Rockies could be successfully crossed in the winter and that train tracks could be built and maintained along this path. This was the route touted by his father-in-law, Missouri senator Thomas Hart Benton, who wanted Congress to designate a central rather than a southern route.

The gold rush and treaties with Mexico and Britain made construction of a transcontinental railroad especially critical in order to bring emigrants to California and Oregon. But Frémont also had very personal goals. He needed to rebuild his crumbling reputation and poor financing. As he explained, "I have a natural desire to do something in the finishing up a great work in which I had been so long engaged."[12]

Frémont had grand plans to write a best-selling and profitable book that would describe in prose and illustrations his years of work surveying the magnificent West. By hiring Carvalho, he hoped to obtain images of what he expected to be a very successful expedition to reinforce the book's adventurous message. Frémont was interested in daguerreotypes not as the final products but as sources for engravings that could be reproduced in his book.[13] The first report of Carvalho's expedition work came in a letter written by Jessie Benton Frémont, who conveyed her husband's sentiments when she wrote: "The party worked well. The Daguerre failed at first but the artist put his heart in his work, & each day Mr. Frémont writes surpassed the work of the day previous. So that he will have faultless illustration."[14]

Having a photographic record was so important to Frémont that he hired a man known only as Mr. Bomer as a backup in case Carvalho did not reach St. Louis in time. When both photographers reached Westport, Kansas, a contest was held between the two and their methods. Carvalho used daguerreotype, in which a light-sensitive metal plate was exposed in the camera and developed with mercury fumes, producing a one-time-only positive image; Bomer's method produced a negative that could be used to make paper positives.[15] Carvalho noted that he could produce images more quickly, claiming that it took Bomer all night to produce a photograph. This was not quite the truth, as what Bomer needed to develop his photographs was not a whole night but darkness. Unable to provide a portable darkroom, Frémont left Bomer and his equipment behind, and it became Carvalho's sole task to document the expedition visually for Frémont's future publications.

Some images based on Carvalho's photographs were published in the first volume of Frémont's 1887 *Memoirs of My Life*.[16] However, Frémont never did publish a book of his own on the fifth expedition. Because of his 1856 presidential nomination and the cancellation of the publishing agreement for the second volume of his memoirs, Frémont did not have the time or the inclination to write.

We know about the expedition only through the pens of Carvalho and James Milligan, a muleteer. Frémont, typical of most explorers, prohibited expedition members from keeping journals; fortunately for historians, Carvalho and Milligan disobeyed. Milligan kept a diary and sent letters— some of his reflections reached a St. Louis newspaper while the expedition was still in Kansas, enraging Frémont.[17] Milligan plainly disliked the Pathfinder, whom he viewed as "domineering and selfish."[18] Frémont could not proceed with a rebellious expedition member, and Milligan did not want to proceed. Complaining of sore feet, Milligan left the expedition at Bent's Fort (present-day eastern Colorado).[19]

Milligan's journal is a valuable record of Carvalho's methods and technical challenges and demonstrates how Carvalho was a fully integrated expedition member. His journal entries parallel some of Carvalho's descriptions and give validity to Carvalho's text. In places, Milligan's journal provides insight into how Carvalho constructed his own memoir. Several of Milligan's entries describe helping Carvalho with his equipment. On November 8 Milligan noted, "Stopped and got out the Daguerreotype apparatus and sent out the Hunters after a Herd of Buffalo in sight to take a picture of the Hunt[.] Washington run an old Bull out of the Herd within 10 yards of the Camera. Seignor Carvalho the artist beat a retreat."[20] In his version of the incident, Carvalho tells not of his retreat but of problems trying to photograph bison in motion.[21] Built for detailed still photography, Carvalho's daguerre camera could not capture the motion of the bison or near or distant subjects with similar clarity.[22] Milligan also commented on the limitations of the camera on windy days.[23] However, the camera performed well on November 9, when Carvalho "[t]ook a Daguerreotype view of the country," and on November 20, when Carvalho photographed a Cheyenne village.[24] Carvalho also described this scene in some detail. Milligan's diary entries were brief and not written for publication, while most of Carvalho's narrative was polished and full of adventure.

While Carvalho tells many stories about the journey, he conceals his Jewish faith, revealing it only to readers who are familiar with Jewish practice. Carvalho was hesitant to eat animals that were not kosher or whose meat reminded him of prohibited or unkosher food. When a porcupine was brought in by the Delaware hunters, he did not eat it, as "[t]he meat . . . looked very much like pork."[25] However, he must have known when he chose to follow Frémont into the wilderness that it would be difficult to observe kosher laws. He did eat meat from animals that were not killed in accordance with kosher ritual in which a knife is used and the blood is drained. Following the Biblical directive, he would not eat a coyote, as carrion-eaters are forbidden. When Carvalho wrote that those who did eat the coyote "were all taken with cramps and vomiting" (114), he implies that their illness validates the wisdom of kosher laws. Carvalho, however, was not strict in his practice. When the expedition was near starvation in the extreme cold of the Rockies and was forced to eat their horses and mules, he ate the "strange and forbidden food" (113), having given up hope of finding game, such as deer, which could be kosher if properly slaughtered. When he did eat horsemeat, he did not eat the blood, which is never kosher (125).

One expedition member succumbed to starvation and frostbite; his death occurred the same day the expedition was rescued and taken to the Mormon town of Parowan. An observer noted in horror that the expedition was in "a state of starvation."[26] Carvalho and Frederick von Egloffstein, the topographer, were quite ill, malnourished, and extremely weak. In no condition to travel farther, they stayed behind when Frémont and most of the original expedition members moved on toward California. Never one to admit even partial failure, Frémont wrote to his father-in-law Senator Benton that the expedition was in "general good health and [had accomplished] reasonable success in the object of our expedition."[27]

Carvalho's experience with Frémont ends here, but his narrative, describing his experiences in Utah and subsequent travels to California, continues. In the mid-nineteenth century, much about the Mormon lifestyle was controversial yet especially intriguing.

Carvalho recuperated in Parowan and then enjoyed the social life of Salt Lake City. Upon regaining his health, he briefly resumed his occupation as a portrait artist and observed the Mormon community by taking an active interest in their religious and family practices. A highlight of his time in Utah was his friendship with Brigham Young, who sat for portraits and invited Carvalho to social functions. This association was typical for Young, as he had a continuing interest in Jews and Judaism. It is obvious in *Incidents of Travel* that Young knew Carvalho's background and made no effort to proselytize him.[28]

Wanting to follow Frémont's 1843 footsteps along the southern trail to illustrate the route and then return home by ship, Carvalho joined a company of Mormons traveling to California. He settled for a short time in Los Angeles and acquainted himself with its Jewish community. Approximately thirty Jews lived in Los Angeles when Carvalho arrived. Most were young merchants, men who sold essential goods and lived above their shops and close to each other. Two of these men, Joseph and Samuel K. Labatt, while they may not have been known to Carvalho, were familiar names to his family. Carvalho's journal calls them "strangers" and does not note that they are Jewish, although the Labatt brothers were the sons of Abraham Cohen Labatt, who, with Carvalho's father, was one of the forty-three men who signed the constitution of the Reformed Society of Israelites on February 15, 1825 in Charleston, when Solomon was ten years old.²⁹ The Labatt family and Carvalho also shared a friendship with Rabbi Julius Eckman, who had served the Charleston community and in 1854 became the first rabbi of Congregation Emanu-El in San Francisco, where the Labatts' father was president and their brother secretary.³⁰ Carvalho was a strong supporter of Eckman, and they had a sympathetic correspondence.³¹ With all this in common, it is curious that Carvalho calls the Labatts strangers, although it fits with his pattern of not overtly divulging his Jewish identity.

While in Los Angeles Carvalho completed portraits of renowned Californians in a painting and daguerreotype studio above the Labatts' Tienda de China on Main Street.³² All three were American-born Sephardic Jews with prior histories of founding and supporting Jewish organizations. Needing a society to purchase land for a cemetery, in July 1854 the trio joined with twenty-eight others in forming the first Jewish organization in the city, the Hebrew Benevolent Society. Samuel Labatt was elected president, and Carvalho became an honorary member. Even though he lived in California only a short time, his presence is recalled in the records of the Los Angeles Jewish community, whose members "[r]esolved unanimously, that the thanks of this meeting be tendered to Mr. S. N. Carvalho for his valuable service in organizing this society, and that he be elected an honorary member."³³ This resolution and the formation of the society was reported locally in the *Los Angeles Star* and nationally to the Jewish community in the *Occident*.

While the founding of the first Jewish organization in Los Angeles was significant in the development of West Coast Jewry, Carvalho did not record this or any observations about Jews in California in his published journal. After more than three months in California, Carvalho started his journey home by way of San Francisco and a ship to the East Coast. The expedition that was to take about six months had kept him away from his home and family for more than a year.

Frémont's presidential election campaign started soon after Carvalho reached his Baltimore home. Actively campaigning for Frémont, Carvalho spoke on street corners and tried to cajole others to join him.[34] At a New York rally for Frémont he was greeted with "three cheers for 'Carvalho,' the Frémont Expedition Artist, Hip! Hip! Hurray!"[35]

Beyond addressing crowds, Carvalho contributed to John Bigelow's 1856 campaign biography of Frémont, which was published in New York by Jackson and Derby. Dedicating thirteen pages to Carvalho's unpolished brief account of the fifth expedition, Bigelow stated that he quoted Carvalho's journal and letters verbatim. Because of this short travelogue, it's likely that Jackson and Derby contracted with Carvalho to write a full narrative of the expedition. In the Bigelow biography, Carvalho provides little praise or other comments about Frémont. However, he added many words of praise for Frémont to the full manuscript, at one point suggesting that if the railroad was built along their route a statue of Frémont should be built at the summit of the Huerfano Butte (Colorado) with "his right hand pointing to California, the land he conquered" (76). Elsewhere, Carvalho suggests that the expedition was in complete agreement that Frémont would make a stellar president (38–39). Despite Carvalho's enthusiastic support, Frémont was defeated.

Carvalho returned to the arts, photography, and painting (western landscapes and portraits, including likenesses of Isaac Leeser and, most notably, Abraham Lincoln).[36] He supported his wife and five children through his engineering inventions, receiving patents for improvements to steam-heating boiler systems that were used in factories and mills.[37]

Carvalho continued his Jewish community work in Baltimore, Philadelphia, and New York. Because of high Jewish immigration rates from central Europe, the German language was becoming dominant in many of the country's elite synagogues. This was problematic for Carvalho, who believed in English-language worship and children's education. In Baltimore, he led the formation of Beth Israel, a Sephardic synagogue that introduced English hymns and prayer along with the traditional Hebrew. Locally, he supported the work of his wife, Sarah, in an English-language Jewish Sunday school.[38] Nationally, he joined the American Jewish community, speaking out strongly for Jews who needed protection in Europe. He served on committees protesting both the exclusion of Jews from some Swiss cantons and the kidnapping by the Papal State of an Italian Jewish child, thought to have been baptized Catholic. Continuing to write both secular and religious works, in 1874 Carvalho wrote a description of a trip to Martinique that was published in *Harper's Monthly*. He later wrote an ambitious philosophical work, "The Two Creations: A Scientific, Hermeneutic and Etymologic Treatise on the Mosaic Cosmogony from the Original Hebrew Tongue," which attempted to reconcile scientific and Biblical creation theories. Carvalho died in 1897 at the age of eighty-two, seven years after John Charles Frémont.

Although Carvalho died more than a hundred years ago, mysteries still surround *Incidents of Travel* and the daguerreotypes. As discussed earlier, Frémont had a strict rule against expedition members keeping journals and publishing accounts; he wanted the final word on his expeditions. It is obvious that Carvalho kept a journal in Utah, and he could have published it, the Mormon appendix, and the illustrations he made on that part of the journey without conflict with Frémont. When Carvalho was able to send letters to his wife, he told her to keep them "from public eye (47)."

Carvalho's own writings do not resolve the issue of whether he had always intended to publish his own account. On his return to the East, he told H. H. Snelling, editor of the *Photographic and Fine Art Journal*, that he did not take "any private notes" on the expedition.[39] However, while in Los Angeles Carvalho made reference to his plans to publish. An article in the *Los Angeles Star* notes that he "intends to publish . . . his journal of the route as soon as he returns home."[40] What is unclear from this quotation is whether he planned to publish an account of his entire trip or just from the time he left Frémont.

Historian Mary Lee Spence suggests that Carvalho would have been prevented from publishing if the 1856 presidential campaign had not kept Frémont from writing his own account.[41] Carvalho's book became a tool of the campaign. Possibly one of Carvalho's goals in campaigning for Frémont and publishing the manuscript was to gain favor and receive his patronage if the electorate sent Frémont to the White House. In September 1856 one of Carvalho's nephews wrote that his uncle had returned home with an unfavorable opinion of Frémont, only to have a change of heart as the election approached.[42] Carvalho may have circumvented Frémont's rule by writing a journal in letter form until he left the expedition, and by letting his book become campaign literature, he could assure its publication and profit from the fame that it would bring him.

For it brought only fame, not fortune. Carvalho received only three hundred dollars from Derby and Jackson for the manuscript. Although offered royalties of five cents a book, he instead chose a three-hundred-dollar lump-sum payment, an unwise choice, as the book was reprinted four times before 1860.[43] The first 1856 printing was sold in New York and London and is extremely rare today.[44] It was also reprinted in the twentieth century: a 1954 edition by the Jewish Publication Society was enhanced with an introduction by the renowned historian Bertram Wallace Korn.[45]

The 1858 edition included a dedication to Jessie Benton Frémont that was not in the two previous editions.[46] According to Carvalho's nephew, Mrs. Frémont gave permission for the dedication in 1856, so it is unclear why it did not appear until two years later.[47] Spence suggests that the dedication was a thank you to Mrs. Frémont for convincing her husband to let Carvalho publish the book. Or was Carvalho trying to win Mrs. Frémont's favor to have use of the daguerreotypes? Carvalho's images did not illustrate any of his own publication.[48]

Most believe that either Carvalho or Frémont carried the daguerreotypes back to the East. When the equipment was left in the mountains, the daguerreotypes were probably not abandoned with it. Carvalho stated that he had "views illustrating the whole country" with him in Los Angeles (at that time "views" usually meant daguerreotypes).[49] On the other hand, Jessie Frémont wrote that the daguerreotypes' "long journeying by mule through storms and snows across the Sierras, then the searching tropical damp of the sea voyage back across the Isthmus, left them unharmed and surprisingly clear, and, [that] so far as is known, [they] give the first connected series of views by daguerre of an unknown country, in pictures as truthful as they are beautiful."[50]

When the daguerrotypes did reach the East, Frémont hired Matthew Brady to make plates. But after the plates were made, their fate as well as the whereabouts of the daguerreotypes is unclear. Some believe they were stored in a warehouse and were burned in a fire during Frémont's term as governor of Arizona. A daguerreotype of an Indian village found in the Brady collection at the Smithsonian Institution may be the only surviving expedition image.

CONCLUSION

> I succeeded beyond my utmost expectations . . . on the summits of the highest peaks of the Rocky Mountains . . . often standing to my waist in snow, buffing, coating, and mercurializing plates in the open air.
> —S. N. Carvalho[51]

Carvalho's writings are significant for the history of the American West, American Jewish life, and the new technology of field photography. Since most expedition narratives were written by the explorers themselves, Carvalho's writings provide one of the few accounts of the inner workings of an expedition. From his vantage point inside the expedition party, Carvalho was able to describe the complicated relationships between expedition members, American Indians, mountain men, and western settlers, and to show the power of the unpredictable environment.

Beyond the expedition, Carvalho critiqued Mormon society and family life, a subject that few Jews chronicled. Continuing a pattern of community building he had established in the East, Carvalho became active in the Los Angeles Jewish community. However, he does not write about this or his other Jewish activities. There may be several reasons for his silence: He could have assumed that his identity was well known and did not need to be stressed; he could have believed that it would hurt Frémont to be identified with a Jew; perhaps Carvalho or Derby and Jackson thought the book would not sell if it was too Jewish; or he may have viewed his religion as a personal matter and not relevant to the published journal. None of these reasons are satisfying, however, and Carvalho's motivations remain unknown.

Finally, as a pioneer of outdoor photography, his determination and skill helped him complete his task. Soon, daguerreotypes became a dated technology, and Carvalho's reputation rested more on his book than on his little-known photographs. In the future more of Carvalho's images may be found, hidden away in public or private collections. But for now it is the book that is the important record; all who are fascinated with the nineteenth-century West will appreciate and enjoy Carvalho's efforts.

Thank you to readers Tom Chaffin, Ruth Haber, Mitchell Richman, and Jonathan D. Sarna. This introduction is better because of your knowledgeable suggestions and comments.

NOTES

1. Brought to Charleston by his brother, a Jewish educator, author, and cantor, David Carvalho arrived in time to volunteer to

defend the city against the British in the War of 1812. In 1824, he became one of the founders and leaders of the short-lived Reformed Society of Israelites, the first American Jewish congregation to introduce reforms in traditional Judaism. See Robert Liberles, "Conflict over Reforms: The Case of Congregation Beth Elohim, Charleston, South Carolina," in *The American Synagogue*, ed. Jack Wertheimer (Cambridge: Cambridge University Press, 1987) and Gary Phillip Zola, *Isaac Harby of Charleston 1788–1828: Jewish Reformer and Intellectual* (Tuscaloosa: The University of Alabama Press, 1994).

2. *Occident* 9 (1851): 208; quoted in Elizabeth Kessin Berman, "Solomon Nunes Carvalho: Painter and Prophet," in *Solomon Nunes Carvalho: Painter, Photographer, and Prophet in Nineteenth Century America*, (Baltimore: The Jewish Historical Society of Maryland, 1989), 10.

3. See Isaac Leeser, *Occident* 11 (1853): 502–3.

4. For more on Carvalho and his work during this period see *Solomon Nunes Carvalho: Painter, Photographer, and Prophet in Nineteenth Century America* (Baltimore: The Jewish Historical Society of Maryland, 1989) and Joan Sturhahn, *Carvalho, Artist-Photographer-Adventurer-Patriot: Portrait of a Forgotten American* (Merrick NY: Richwood Publishing, 1976).

5. Bernard P. Fishman, "Solomon Nunes Carvalho: Photographer" in *Solomon Nunes Carvalho,* 26.

6. Carvalho regularly traveled between Charleston, Philadelphia, Baltimore, and New York and was a known photographer throughout the East. It is likely that Carvalho was in New York in 1853 demonstrating his enameling process. Frémont may have met Carvalho through the artist Samuel F. B. Morse, who knew Sen. Thomas Hart Benton (Frémont's father-in-law) and may have been acquainted with Carvalho. Also, others suggest that Frémont was attracted to Carvalho because they shared a Charleston background. Exactly how they met, however, is unrecorded.

7. Jessie Benton Frémont, "Some Account of the Plates," in, *Memoirs of My Life,* by John C. Frémont (Chicago: Clarke & Company, 1887), xv.

8. For information about field photography see Martha A. Sandweiss, *Print the Legend: Photography and the American West*, (New Haven: Yale University Press, 2002).

9. It is unclear whether Frémont or Carvalho paid for the equip-

ment, most likely Frémont. See Donald Jackson and Mary Lee Spence, ed., *The Expeditions of John Charles Frémont*, vol. 3, *Travels from 1848 to 1854* (Urbana: University of Illinois Press, 1984), liv.

10. Letter from S. N. Carvalho to H. H. Snelling, Esq., March 20, 1855, in *The Photographic and Fine Art Journal* (April 1855), 124-25.

11. For more about the expedition see Patricia Joy Richmond, *Trail to Disaster* (Boulder: University Press of Colorado, 1990) and Tom Chaffin, *Pathfinder: John Charles Frémont and the Course of American Empire* (New York: Hill and Wang, 2002).

12. Jackson and Spence, *Expeditions of Frémont*, 3:381.

13. Sandweiss, *Print the Legend*, 107. Photographers were usually not given credit for their work; rather, the engravers of the final images were acknowledged.

14. Letter from Jessie Benton Frémont to Francis Preston Blair, winter 1853–54, in Pamela Herr and Mary Lee Spence, ed., *The Letters of Jessie Benton Frémont* (Urbana: University of Illinois Press, 1993), 56.

15. Mark J. Stegmaier and David H. Miller, ed., *James F. Milligan: His Journal of Frémont's Fifth Expedition, 1853–1854, His Adventurous Life on Land and Sea* (Glendale CA: The Arthur H. Clark Company, 1988), 104, and Sandweiss, *Print the Legend*, 101.

16. Frémont, *Memoirs*.

17. The complete journal was kept in private hands until its 1988 publication. Stegmaier and Miller, *James F. Milligan*, 126.

18. Stegmaier and Miller, *James F. Milligan*, 136.

19. Stegmaier and Miller, *James F. Milligan*, 138. Frémont ordered Milligan to remain at the fort to care for expedition animals until his expected return in May.

20. Stegmaier and Miller, *James F. Milligan*, 130.

21. Although Carvalho uses "buffalo" in his text, bison is the proper term for the North American species. Solomon Nunes Carvalho, *Incidents of Travel and Adventure in the Far West* (Philadelphia: Jewish Publication Society of American, 1954), 64.

22. Sandweiss, *Print the Legend*, 101.

23. Stegmaier and Miller, *James F. Milligan*, 130.

24. Stegmaier and Miller, *James F. Milligan*, 131, 136.

25. Korn, introduction to *Incidents of Travel*, 170.

26. Stegmaier and Miller, *James F. Milligan*, 93.

27. Jackson and Spence, *Expeditions of Frémont*, 3:469.

28. While there were probably no Jews in Utah while Carvalho was recuperating, Fanny and Julius Brooks arrived by wagon train that fall, and soon there was a growing community. [For more about Fanny and Julius Brooks, see Ava F. Kahn, *Jewish Voices of the California Gold Rush: A Documentary History 1849–1880* (Detroit: Wayne State University Press, 2002). For Jews in Utah see Eileen Hallet Stone, *A Homeland in the West* (Salt Lake City: University of Utah Press, 2001) and Leon L. Watters, *The Pioneer Jews of Utah* (New York: American Jewish Historical Society, 1952).] Young told Mary Goldsmith Prag, the wife of a Salt Lake City merchant, that "no true Jew can be converted to Mormonism." [Ava F. Kahn, ed., *Jewish Life in the American West* (Berkeley: Heyday Books, 2004), 65.] He probably reached this conclusion about Solomon Carvalho.

29. Wertheimer, *American Synagogue*, 277.

30. For more about Carvalho's relationship with Eckman, see Korn, introduction to *Incidents of Travel*, 29.

31. Kahn, *Jewish Voices*, 173

32. For more about Carvalho in Los Angeles see William M. Kramer, "Solomon Nunes Carvalho Helped in Founding the Los Angeles Jewish Community," *Western States Jewish History* 28 (July 1996):4.

33. *Occident* 12 (September 1854): 327.

34. Letter from Carvalho's nephew Ritterband, as quoted in Sturhahn, *Carvalho,* 131.

35. Sturhahn, *Carvalho*, 129.

36. In this portrait painted after Lincoln's assassination, Carvalho honors the martyred president as the "Great Emancipator," a recent theme in 1865. The painting of Lincoln was given to Brandeis University in 1960.

37. Berman, "Solomon Nunes Carvalho," 21n50.

38. Berman, "Solomon Nunes Carvalho," 16.

39. Carvalho to Snelling, *Photographic and Fine Art Journal*, 124–25.

40. "Carvalho the Artist," *Los Angeles Star*, July 8, 1854, 2.

41. Jackson and Spence, *Expeditions of Frémont*, 3:403.

42. Letter of Jacob S. Ritterband, as quoted in Sturhahn, *Carvalho*, 130.

43. Korn, introduction to *Incidents of Travel*, 37. It was published by Derby and Jackson in 1856, 1857, 1859, and 1860.

44. Jackson and Spence, *Expeditions of Frémont*, 3:403.

45. The Jewish Publication Society published *Incidents of Travel* during a period in which they focused on primary source materials in American Jewish history that were edited for general readers. Jonathan D. Sarna, *JPS: The Americanization of Jewish Culture, 1888–1988* (Philadelphia: Jewish Publication Society, 1989), 255. Other twentieth-century publications of *Incidents of Travel* include Kraus Reprint Co. (New York, 1971) and Arno Press (New York, 1973), a reprint of the 1857 Derby and Jackson.

46. This Bison Books edition is a reproduction of the 1858 edition.

47. Letter from Ritterband, in Sturhahn, *Carvalho*, 130.

48. The frontispiece of the 1858 edition has nothing to do with the fifth expedition. Another mystery.

49. "Carvalho the Artist," *Los Angeles Star*, July 8, 1854, 2.

50. Jessie Benton Frémont in Frémont, *Memoirs*, xvi.

51. Carvalho to Snelling, *Photographic and Fine Art Journal*, 124–25.

TO

MRS. JESSIE BENTON FREMONT,

THIS BOOK

IS, BY PERMISSION, MOST RESPECTFULLY INSCRIBED.

PREFACE.

In preparing this volume for publication, I have not followed any established system of arrangement.

The incidents are most of them transcripts from original letters, written in the familiar style of friendly correspondence.

The description of a journey from Great Salt Lake City to San Bernandino, is an exact copy from my journal, written after many days of wearisome travel.

The Mormon Episodes, I have rendered almost verbatim from personal relations by the parties themselves, and not from "hearsay."

While the Latter-day Saints publicly adopt every opportunity to openly avow and zealously propagate the System of Polygamy—in direct opposition to the established and acknowledged code of morality, as practised by all civilized nations—I but exercise my prerogative in exposing some of its abuses, which I consider destructive to morality, female delicacy, and the sanctity of marriage.

To prove the correctness and authenticity of my statements, with regard to the moral and ecclesiastical views of the Latter-day Saints, I have appended to the end of this volume, several discourses and addresses, some of which were delivered during my sojourn in Utah, by President Brigham Young and his apostles, and reported by G. D. Watt, Esq. of Great Salt Lake City

For the rest, I submit myself to an indulgent public.

THE AUTHOR.

BALTIMORE, MD., *September*, 1856.

CONTENTS.

CHAPTER I.

First Introduction to Col. Fremont—Author's previous Opinion of Him—His Impulsive Resolution to accompany Col. Fremont, as Artist of the Exploring Expedition across the Rocky Mountains, ... 17

CHAPTER II.

Preparations for the Journey—Daguerreotype Outfit—Scientific Knowledge required—Departure from New York—Alden's Preserved Food—Esteem of Col. Fremont's former Companions for him—Arrival at St. Louis—Steamboat F. X. Aubrey—Compagnons de Voyage—Arrival at Kansas, ... 20

CHAPTER III.

Landing of Camp Equipage—Westport—First Camp Ground—Preparations—Extortion—Author and Companions—First Daguerreotypes—Rain Storm—Distribution of Arms and Ammunition—Engagement of Delaware Chiefs—Branding of Animals—California Saddle-Horses—Selects his Pony—Becomes his own Ostler—Description of Catching a Mule on the Mountains—Examination of Camp Equipage—Trial Start—First Camp, ... 23

CHAPTER IV.

Shawnee Mission—Indisposition of Col. Fremont—He returns to Westport—The Expedition proceeds onward—Sunrise on the Prairies—Meeting of our Delawares—Pottawatomie trading Post—Author turns Carpenter—Expediency—A Kansas Blacksmith—"Astonishment"—Quarter Master—Persecution—Success against Conspiracy. 29

CHAPTER V.

Letter to W. H. Palmer—Col. Fremont's Return to St. Louis, and increased Illness—Expedition under charge of Delawares—Camp Proceeds to "Smoky Hills"—Fort Riley—Solomon's Fork—First Buffalo—Barometers go on a Buffalo Hunt—Encampment on "Salt Creek"—Indian Method of Cooking Buffalo Meat—Olla Podrida—Wasting of Provisions—Kinnikinick—Havana Segars—Indian Amusements—Camp Life—Hewers of Wood and Drawers of Water—Author's Opinion of Col. Fremont—He Nominates him for the Presidency, ... 34

CHAPTER VI.

Kansas Territory—Arkansas and Kansas Rivers—Tributaries—Timber—River Bottoms—Varieties of Game—Cereals—Coal—Geological Formation—Adventure in the Woods—Wild Grapes—Indian Method of procuring them—Brandy vs. Poison—Return of the Author's Brandy-flask—He turns Washerwoman—Novel Mode of Mangling Clothes—Lost Mule—Beaver Trappers—Rifle Practice, ... 40

CHAPTER VII.

Delaware Medicine Man—Illness of Capt. Wolff—Author turns Doctor—Empty Commissariat—Expedition to Fort Riley for Fresh Supplies—Professor Espy's Theory of Rain—Indians on Kansas Prairies—Sleet and Snow Storm—Tent Blown Down—Approach of Cold Weather—"Pony Missing," 46

CHAPTER VIII.

Author's First Buffalo Hunt—Pursuit—Perilous Situation—Mode of Attack by the Indians—Solitary and Alone—Pony killed for Food—An Ill Wind that blows Nobody any Good—Incredulity of Indian Hunters—Return to Camp—Prairies on Fire—Suffocating Smoke—Mr. Egloffstein on a Wolf Hunt—Fire Light and Moonlight—Camp surrounded by Fire—Dangerous Situation—Arrival of Colonel Fremont—Preparations to resume our Journey—Escape through the Blazing Element, . 50

CHAPTER IX.

Walnut Creek—Necessity of strict discipline—Neglect of duty—Horses stolen—Cheyenne Indians—Thieves overtaken—Watchfulness of Col. Fremont—Immense herds of Buffalo—Buffalo hunt on a large scale—Buffalo chips—Prairie dogs, Owls, &c.—Indians in camp—Raw Antelope liver, ... 62

CHAPTER X.

Cheyenne Indian Village—"Big Timber"—Daguerreotypes—Indian Papoose—Author is Suddenly Changed into a Magician—Silver and Brass Bracelets—Portrait of Indian Princess—"*Presto, pass!*"—Moccasins—Cheyennes and Pawnees at War—Grand Scalp-dance—Col. Fremont a Spectator—Dinner with the Chief—Rosewood Carved Furniture not in Use—Duties of Indian Women—Employment of the Men, . 67

CHAPTER XI.

Bent's Trading Post—Purchase Fresh Animals—Buffalo Robes—Immense Lodge—Fremont's Lodge—Doctor Ober—His Scientific Knowledge—Attachment of the Author to him—His Preparation to return to the States—Arkansas River—Giant Cotton Woods—Islands in the Arkansas—Bent's Fort destroyed by Indians—Preparations to cross the Mountains—First View of the Rocky Mountains—Bid adieu to Doctor Ober, ... 71

CONTENTS.

CHAPTER XII.

Journey up the Arkansas—Bent's Fort—Huerfano River and Valley—Description of the Country—Huerfano Butte—Behind Camp—Daguerreotypes—Scientific Observations—Approach of Night—Trail Lost, and Encampment in the Woods—Buffalo Robes and Blankets—Col. Fremont sends to find us—Bear Hunt—Roubidoux Pass—Emotion of Col. Fremont when Looking upon the Scene of his Terrible Disaster on a Former Expedition—Found a Half Starved Mexican—Col. Fremont's Humanity—His Skill in Pistol Shooting, ... 75

CHAPTER XIII.

Sand-hill Pass—San Louis Valley—Natural Deer-park—Smoked Venison—Last sight of Game—Rio Grande del Norte—Sarawatch—Cochotope Pass—First Snow in Mountains—Gunnison's Wagon Trail—Summit of Pass—Waters commence to flow towards the Pacific—Encampment—Immense Rugged Mountain—Impracticability of ascent by Mules—The Author ascends on Foot—Col. Fremont accompanies him—Daguerreotype Panorama from its Summit—Col. Fremont's Consideration for his Men—Sublimity—First View of Grand River—Reflections—Return to Camp, . . 80

CHAPTER XIV.

Intense Cold—Author's First Journey on Foot—Immense Mountains of Snow—Escape of his Pony—Lose Sight of Companions—Arrival at top of the Mountain—Pony Recovered—Revolution of Feeling—Indian Gratitude Exemplified—Horse Steaks Fried in Tallow Candles—Blanc Mange—New Year's Day—Dangerous Ascent of a Mountain—Mules tumble Down—Animals Killed—Successful Attempt Next Day—Camp in four feet of Snow—Coldest Night—Sleep out in open Snow, . . 84

CHAPTER XV.

Descent of Snow Mountains—Gun for a Walking-stick—Indian Tracks—Examination of Arms—Predicament of the Author—Lecture from Col. Fremont—Wild Horse Killed by Indians—Utah Indian Village—Encampment—Trade for Venison—Camp at Night Surrounded by armed Indians—They Demand Payment for the Horse Killed by the Indians—Col. Fremont's Justice—Indians want Gunpowder—Their Demand Refused—Massacre of the Party Threatened—Defiance—Pacification—Author Trades for a Horse—He Leaves his Colt's Revolver in Camp—Runaway Horse—Author Finds himself in a Sage Bush—Pistol Recovered—Trouble in Perspective—Exchanges Horses—Lame Horse—Author on Foot—Regrets that he was not Educated for a Horse-Breaker, ... 89

CHAPTER XVI.

Grand River—Descent of Mounted Indians into Camp—Military Reception—Their demands—Trouble Expected—Excitement of the Author—Exhibition of Colt's Revolvers—Col. Fremont's Knowledge of Indian Character—The Great Captain in his Lodge—Alarm of the Indians—Quadruple Guard—Departure of Indians—Vigilance the price of Safety—Crossing of the Grand River—Horse Killed for Food—Review of Our Position—Impressive Scene—Cold Night—Mr. Fuller—Whites without Food—Beaver Shot—The Camp under Arms—False Alarm, ... 96

CONTENTS.

CHAPTER XVII.

Divide between Grand and Green River—Capt. Gunnison's Trail—Without Water—Formation of the Country—Castellated Bluffs—Green River Indians—Crossing of the Green River—Interview with Indians—Disappointment—Grass-seed—Manner of Preparing it for Food—Horse Purchased—Starving Condition of the Whites—Incident Exhibiting the Moral Dishonesty of one of the Men—Name not Published—Dinner on Porcupine—"Living Graves"—Tempestuous Night—Reflections on Guard—No Grass—Frozen Horse Liver—Blunted Feelings, ... 104

CHAPTER XVIII.

Careless Packing of Animals—Mule Missing—Their value as Roadsters—Col. Fremont's Horse gives out—His Humanity Exemplified—Wolf killed for Food—Raven Shot—River Bottom—Original Forest—Large Camp Fires—Terrible Rain Storm—Disagreeable Bed—Darkness—Fires Extinguished—Value of Rain—Glorious Sunrise—Contrast with Home Comforts, ... 111

CHAPTER XIX.

Crippled Condition of the Party—Mr. Oliver Fuller—Mr. Egloffstien—Mr. Fuller gives out—His Inability to Proceed—Mr. Egloffstien and the Author continue on to Camp for Assistance—Col. Fremont sends Frank Dixon after him—Sorrow of the Camp—Mr. Fuller's Non-Appearance—Delawares sent out to Bring the Men in—Return of Frank Almost Frozen—Restoration of Mr. Fuller—Joy of the Men—Serious Thoughts—The Author Prepared to Remain on the Road—His Miraculous Escape, . 117

CHAPTER XX.

Author nearly gives Out—Family Portraits—Fresh Courage—Dangerous Situation—Lonely Journey—Darkness—Snow Storm—Arrival at Camp—"Col. Fremont's Tent"—Interview with Col. Fremont—"Cache"—Men on Foot—Daguerreotype Apparatus buried in the Snow—Sperm Candles—Men Mounted on Baggage Animals—Seveir River Beaver Dams—Modus Operandi of killing Horses for Food—Entrail Soup—Hide and Bones Roasted—Influence of Privation on Human Passions, 122

CHAPTER XXI.

Unsuccessful Attempt to Force a Passage in the Mountains—Delawares sent out to Explore—Their Return—Col. Fremont, Capt. Wolff, and Solomon in Council—Unfavorable Report of Capt. Wolff—Col. Fremont's Determination—Astronomical Observations at Midnight—Col. Fremont's Correctness and Skill Illustrated—Tremendous Mountains of Snow—Successful Ascent on Foot, without Shoes or Moccasins—Tribute to the Genius of Fremont—Col. Fremont's Lodge at Meal-Time—Mr. Oliver Fuller's Death—Sorrow of his Companions—His Last Hours—His Virtues—Indian Camp—Arrival at Parowan—Burial of Mr. Fuller—Author's Physical Condition—Mormon Sympathies—Mr. Heap and his Wives—Mormon Hospitality, . 128

CHAPTER XXII.

Sojourn at Parowan—Colonel Fremont refits his Expedition—Illness of the Author—His Inabillity to Proceed—Takes Leave of Col. Fremont—Mr. Egloffstien and the Author leave to go to Great Salt Lake City in a Wagon—Col. Fremont's Departure—

CONTENTS. xxxiii

Mormons for Conference—Arrival at Salt Lake City—Massacre of Capt. Gunnison—Interview with Lieut. Beckwith—Mr. Egloffstien appointed Topographical Engineer—Painting Materials—Kinkead and Livingston—Brigham Young—Governor's Residence—Apology for Mormonism among the Masses—Their previous Ignorance of the Practice of Polygamy, ... 139

CHAPTER XXIII.

Governor Brigham Young—Author's Views on Polygamy—Baptismal Ceremony—Doctrines and Covenants, ... 146

CHAPTER XXIV.

Grand Ball at Salt Lake City—Etiquette—Culinary Preparations—Cost of Entertainment—Author opens the Ball with one of the Wives of the Governor—Beautiful Women—Waltzing and Polkas Prohibited—Mrs. Wheelock—The "Three Graces"—Extraordinary Cotillion—Mormon Wedding—Spiritual Wives—Favorable Impression of the Public Social Life of the Mormons, ... 155

CHAPTER XXV.

"Golightly"—His Occupation and Character—Author Patronizes Him—Mrs. Golightly—She thinks Shakspeare did not understand the Passions of Men—"Oh! Frailty, thy Name is Man!"—Affecting Incident, ... 160

CHAPTER XXVI.

Extraordinary Abuses of the Spiritual Wife System—Fanny Littlemore—The Writer paints her Portrait—Her early Life—Attempt by her Parents to force her to marry her Uncle at Nauvoo—Her Escape to St. Louis—She writes to her Lover—Terry Littlemore—Marriage—Extraordinary Letter—Fanny's Mother exchanges Husbands with her Aunt—Her Father also exchanges Wives with her Uncle—Fanny's journey to Salt Lake—Terry Littlemore becomes a Mormon—Fanny opposed to Mormonism—Her two Sisters become spiritual Wives of a distinguished Mormon—She meets her Father and Mother in Salt Lake—The Writer becomes acquainted with her Mother and Uncle—His Journey to Parowan with them—Verification, ... 166

CHAPTER XXVII.

Arrival of the California Mail—Murder of Mr. Lamphere by Indians on Santa Clara—Hot Springs—Singular Phenomenon—Hot and Cold Springs—Mica—Sulphur—Plumbago—Rock Salt—Death of Willard B. Richards—Heber C. Kemball—Welsh Colony—Lieut. Beckwith's Departure for California, ... 175

CHAPTER XXVIII.

Departure from Great Salt Lake City—Equipments for the Journey—Author Paints Portraits of Gov. Young and Apostles—His Restoration to Health—Snow Storm—Cotton Wood Settlement—Willow Creek—Lehigh—Utah Lake—Snow Storm—Pleasant Grove—Provost—Payson, ... 180

CONTENTS.

CHAPTER XXIX.

Join Governor Young and Parley Pratt—Hospitality of the Mormons—Apostle Benson—Petetnit—Nephi—Wakara (Indian Chief)—Wakara's Camp Ground—Brigham Young's Wife—Long Caravan—Arrival at Wakara's Camp—His Refusal to meet the Governor—Treaty of Peace not Concluded—Presents of Cattle, etc., to Wakara—Grand Council of Indians and Mormons—Speech of an Old Chief—Address of a "San Pete Chief"—Wakara Refuses to Speak—He Dissolves the Council—Reassembling of the Council—Brigham Young's Address—Speech of "Wakara"—Peace Proclaimed—Calumet Smoked—Indian Capture of Children—Brigham Young's Residence, . . 185

CHAPTER XXX.

Portrait of Wakara—Indian chiefs, to accompany the Expedition to Harmony City—Seveir River—Swollen Waters—Wagons ferried over—Col. Fremont—Fillmore City—Massacre of Capt. Gunnison—Parowan Indians—Kanosh—Capt. Morris—His conduct justified—Author trades for a Horse—Extraordinary Phenomenon of Insects, 195

CHAPTER XXXI.

Corn Creek—Meadow Creek—Exploration of Vinegar Lake—Mephitic Gas—Sulphuric Acid—Sulphur—Alum—Volcanic Appearance of the Country—Beaver River Valley—Lieut. Beale's Pass into the Valley of the Parowan—Col. Fremont's Pass in the same Valley—Author crosses his own Trail made three Months before—His Feelings on the Occasion—Red Creek Cañon—Hieroglyphics—Granite Rocks—Remains of a Town—Arrival at Parowan—Brigham Young—Old Acquaintances, 202

CHAPTER XXXII.

Description of Parowan—Cedar City—Fish Lake—Iron Ore—Bituminous Coal—Future Destiny of Cedar City—Henry Lunt—Affecting Incident—Portrait of a dead Child—A Mother's Gratitude—Harmony City—Parley Pratt—Piede Indians—Personal Privations of Mormons—Bid Adieu to Gov. Young—Letter of Introduction to President of San Bernardino, ... 209

CHAPTER XXXIII.

On the Road to California—Iron Springs—Meadow Springs—Entrance to Las Vegas de Santa Clara—Prairie Flowers—Rim of the Basin—Santa Clara River—Difficulty of Crossing with Wagons—Wounded Indian—Serpentine Course of the River—Waterfall—Natural Cave, ... 215

CHAPTER XXXIV.

Romantic Pass—Rio Virgin Valley—Sterile County—River Bottoms—Acacia Groves—Abrupt Descent—Formation of the Country—Pah Utahs—Indian Bow and Arrows—Orange color Berries—Effect on the System—Digger Indians—Baptized into Mormon Faith—Steep descent—Divide between Rio Virgin and Muddy Rivers—Difficult travelling—Muddy River described—Author lends his Horse—Approach to the "Great Desert,". 220

CONTENTS.

CHAPTER XXXV.

Preparations to Cross the Jornada—Fifty-five Miles without Water or Grass—Deserted Wagons on the Road—Dead Oxen and Mules—Emigrant Party—Clouds of Dust—Oasis—Delicious Water—Extraordinary Fresh Water Buoyant Spring—Impossibility for a Man to sink in it—Never before Described—Another Jornada of Forty Miles—Col. Reese's Train—Detention—Reese Cut off—Snow-Capped Mountains—Bad Roads—Mineral Springs—My Mule in Harness—Animals giving out, ... 228

CHAPTER XXXVI.

Peg-leg Smith—Gold Explorers—Enter upon the Desert—Road strewn with Dead Oxen—Poisoned Atmosphere—Deserted Wagons and Horses—Howling Wilderness—Excessive Heat—Bitter Springs—Polluted by Dead Animals—Bunch Grass—Reflections—Mohahve River—Deserts Surmounted—Horses give Out—On Foot—Dig for Water in the Sand—Pleasant Weather—Snowy Mountains—Crossing of the Mohahve River—Agave Americana—Cajon pass Sierra Nevada—Descent into the Valley of San Bernandino—Arrival at San Bernandino—Variations of the Compass, . 234

CHAPTER XXXVII.

Journey to Los Angeles—Catholic Missions—Fields of Mustard—California Ladies—Morals of the People—Gamblers—Description of a "Hell"—Climate of Los Angeles—Delicious Fruit—California Wine—Don Manuel Domingues—Rancho—Menada—Breaking a Horse—Portraits of Domingues—Salt Lake—Asphaltum Lake—Hot Springs of San Juan de Campestrano—Analysis—Geological Examination—Remains of a Mastodon—Don Pio Pico—Ground Squirrels—Strychnine—Brothers Labatt—Their Example worthy to be Imitated, ... 242

SPIRITUAL WIFE SYSTEM.

A Revelation on the Patriarchal order of Matrimony, or Plurality of Wives. Given to Joseph Smith, the Seer, in Nauvoo, July 12th, 1843, ... 1

CELESTIAL MARRIAGE.

A Discourse delivered in the Tabernacle, Great Salt Lake City, ... 12

INDIAN HOSTILITIES AND TREACHERY.

Consequences of Obedience and Disobedience—Policy towards the Indians—Vigilance. An Address delivered at the Tabernacle, Great Salt Lake City, by Brigham Young, 39

USE AND ABUSE OF BLESSINGS.

An Address delivered by Brigham Young, at the Tabernacle, Great Salt Lake City, 61

MORMONISM.

A Discourse delivered by Parley P. Pratt, in the Tabernacle, Great Salt Lake City, 79

LEGITIMACY AND ILLEGITIMACY.

A Sermon delivered at the General Conference, Tabernacle, Great Salt Lake City, 104

CARVALHO'S TRAVELS AND ADVENTURES.

CHAPTER I.

First Introduction to Col. Fremont—Author's previous Opinion of Him—His Impulsive Resolution to accompany Col. Fremont, as Artist of the Exploring Expedition across the Rocky Mountains.

ON the 22d August, 1853, after a short interview with Col. J. C. Fremont, I accepted his invitation to accompany him as artist of an Exploring Expedition across the Rocky Mountains. A half hour previously, if anybody had suggested to me, the probability of my undertaking an overland journey to California, even over the emigrant route, I should have replied there were no inducements sufficiently powerful to have tempted me. Yet, in this instance, I impulsively, without even a consultation with my family, passed my word to join an exploring party, under command of Col. Fremont, over a hitherto untrodden country, in an elevated region, with the full expectation of being exposed to all the inclemencies of an arctic winter. I know of no other man to whom I would have trusted my life, under similar circumstances.

Col. Fremont's former extraordinary explorations, his astronomical and geographical contributions to the useful sciences, and his successful pursuit of them under difficulties, had deeply interested me, and aided in forming for him, in my mind, the beau ideal of all that was chivalrous and noble.

His conquest of California, appointment as Governor by Commodore Stockton, the jealousy and persecution by General Kearney for not acknowledging him instead of Commodore Stockton as commander-in-chief, his court-martial and subsequent finding of the court, are matters of American history, and they reflect no dishonor on the individual who was a distinguished examample of the ingratitude of republics.

The recognition of his claims on the American public by the citizens of Charleston, S. C., who presented him with an elegant sword and golden scabbard, satisfied me that I had formed no incorrect estimate of his character, and made me feel an instinctive pride that I, too, drew my first breath on the same soil that gave birth to heroes and statesmen.

Entertaining these feelings, the dangers and perils of the journey, which Col. Fremont pointed out to me, were entirely obscured by the pleasure I anticipated in accompanying him, and adding my limited skill to facilitate him in the realization of one of the objects of the expedition—which was to obtain an exact description of the face of the country over which we were to travel.

The party consisted of twenty-two persons; among them were ten Delaware chiefs; and two Mexicans. The officers were: Mr. Egloffstein, topographical enneer; Mr. Strobel, assistant; Mr. Oliver Fuller, assistant engineer; Mr. S. N. Carvalho, artist and daguerreotypist; Mr. W. H. Palmer, passenger.

The expedition was fitted out, I think, at the individual expense of Col. Fremont.

CHAPTER II.

Preparations for the Journey—Daguerreotype Outfit—Scientific Knowledge required—Departure from New York—Alden's Preserved Food—Esteem of Col. Fremont's former Companions for him—Arrival at St. Louis—Steamboat F. X. Aubrey—Compagnons de Voyage—Arrival at Kansas.

THE preparations for my journey occupied about ten days, during which time I purchased all the necessary materials for making a panorama of the country, by daguerreotype process, over which we had to pass.

To make daguerreotypes in the open air, in a temperature varying from freezing point to thirty degrees below zero, requires different manipulation from the processes by which pictures are made in a warm room. My professional friends were all of the opinion that the elements would be against my success. Buffing and coating plates, and mercurializing them, on the summit of the Rocky Mountains, standing at times up to one's middle in snow, with no covering above save the arched vault of heaven, seemed to our city friends one of the impossibilities—knowing as they did that iodine will not give out its fumes except at a temperature of 70° to 80° Fahrenheit. I shall not appear egotistical if I say that I encountered many difficulties, but I was well prepared to meet them by having previously acquired a scientific and practical knowledge of the chemicals I used, as well as of the theory of light: a firm determination to succeed also aided me in producing results which, to my knowledge, have never been accomplished under similar circumstances.

While suffering from frozen feet and hands, without food for twenty-four hours, travelling on foot over mountains of snow, I have stopped on the trail, made pictures of the country, re-packed my materials, and found myself frequently with my friend Egloffstien, who generally remained with me to make barometrical observations, and a muleteer, some five or six miles behind camp, which was only reached with great expense of bodily as well as mental suffering. The great secret, however, of my untiring perseverance and continued success, was that my honor was pledged to Col. Frèmont to perform certain duties, and I would rather have died than not have redeemed it. I made pictures up to the very day Col. Fremont found it necessary to bury the whole baggage of the camp, including the daguerreotype apparatus. He has since told me that my success, under the frequent occurrence of what he considered almost insuperable difficulties, merited his unqualified approbation.

I left New York on the 5th September, 1853, having in charge the daguerreotype apparatus, painting materials, and half a dozen cases of Alden's preserved coffee, eggs, cocoa, cream, and milk, which he sent out for the purpose of testing their qualities. There was in them sufficient nourishment to have sustained twenty men for a month. I purchased a ticket by the Illinois River to St. Louis, but the water was so low in the river that it was deemed advisable to cross over to Alton by stage, as I was afraid of being detained. The cases of instruments were very heavy, and the proprietor of the stage refused to take them; it being night, I remonstrated with him, telling him of the importance that they should arrive at St. Louis; he peremptorily refused to take them. I, of course, had to succumb, and remarked inadvertently how disappointed Col. Fremont would be in not receiving them. At the mention of Col. Fremont's name, he asked me if those cases were Fremont's? I told him, yes. He sang out for his boy to harness up an extra team of horses, and stow away the boxes. "I will put them through for Fremont, without a cent expense. I was with him on one of his expeditions, and a nobler specimen of mankind does not live about these parts." I was put through in good time, but he would not receive a cent for my passage, or freight of the boxes, which together would have amounted to eight dollars.

I arrived at St. Louis at twelve o'clock. Col. Fremont was at Col. Brant's house, where I immediately called. The Colonel was very glad to see me; he had telegraphed several times, and I had been anxiously expected. We left that same afternoon in the steamer F. X. Aubrey, for Kansas. On board, I found Mr. Egloffstien, the topographical engineer, Mr. Oliver Fuller, and Mr. Bomar, the photographist. Our journey was somewhat protracted by the shallowness of the water in the river, and we did not arrive at Kansas until the 14th.

CHAPTER III.

Landing of Camp Equipage—Westport—First Camp Ground—Preparations—Extortion—Author and Companions—First Daguerreotypes—Rain Storm—Distribution of Arms and Ammunition—Engagement of Delaware Chiefs—Branding of Animals—California Saddle-Horses—Selects his Pony—Becomes his own Ostler—Description of Catching a Mule on the Mountains—Examination of Camp Equipage—Trial Start—First Camp.

When we landed, we met Mr. Palmer and several of the men who were to accompany the Expedition as muleteers, etc. The equipage of the camp that had been previously shipped from St. Louis, had arrived safely. As soon as our baggage was landed, it, together with the rest of the material, was transported by wagons to camp near Westport, a few miles in the interior.

Our tents were raised, and active preparation for our journey was immediately commenced. Several droves of mules came in next day from which Col. Fremont selected a few. Very near two prices were exacted by the owners; it being necessary that we should proceed without delay, we were obliged to submit to extortion.

Mr. Egloffstien, Mr. Bomar and myself, found comfortable quarters at a hotel where we put up, in order to be ready for the journey, our various apparatus.

Mr. Bomar, proposed to make photographs by the wax process, and several days were consumed in preparing the paper, etc. I was convinced that photographs could not be made by that process as quickly as the occasion required, and told Col. Fremont to have one made from the window of our room, to find out exactly the time. The preparations not being entirely completed, a picture could not be made that day; but on the next, when we were all in camp, Col. Fremont requested that daguerreotypes and photographs should be made. In half an hour from the time the word was given, my daguerreotype was made; but the photograph could not be seen until next day, as it had to remain in water all night, which was absolutely necessary to develop it. Query, where was water to be had on the mountains, with a temperature of 20° below zero? To be certain of a result, even if water could be procured, it was necessary by his process, to wait twelve hours, consequently, every time a picture was to be made, the camp must be delayed twelve hours. Col. Fremont finding that he could not see immediate impressions, concluded not to incur the trouble and expense of transporting the apparatus, left it at Westport, together with the photographer. The whole dependence was now on me. Col. Fremont told me if I had the slightest doubts of succeeding, it were better to say so now, and he would cancel the agreement on my part, and pay me for my time, etc.

On the night of the 20th, all hands slept in camp, a heavy rain-storm drenched us completely, giving to the party an introduction to a life on the prairies. The necessity of India-rubber blankets became evident, and I was dispatched to Westport to procure them. There were none to be had. I sent a man to Independence to purchase two dozen; he travelled thirty miles that night, and by ten next morning I had them in camp. They were the most useful articles we had with us; we placed the India-rubber side on the snow, our buffalo robes on the top of that for a bed, and covered with our blankets, with an India-rubber blanket over the whole—India-rubber side up, to turn the rain. We generally slept double, which added to our comfort, as we communicated warmth to each other, and had the advantage of two sets of coverings. During the whole journey, exposed to the most furious snow-storms, I never slept cold, although when I have been called for guard I often found some difficulty in rising from the weight of snow resting on me.

The distribution of arms and ammunition to the men occupied a portion of the next day. Each person had a rifle and Colt's revolver. Some of the Delawares had horsemen's pistols also. The messenger Col. Fremont sent to the Delaware camp returned, with a number of braves, some of whom had accompanied Col. Fremont on a former expedition—he selected ten, among whom was a chief named Solomon, who had been with him before, and for whom Col. Fremont felt a great friendship. They were entertained with dinner, and after a smoke, each had a small quantity of the brandy we brought for medicinal purposes. They left us, to make preparations for the expedition, and to join us near the Kansas River, about one hundred miles westward.

A most amusing scene, although attended with some pain to the animals, was enacted to-day; it was the process of branding them with a distinctive mark. We had an iron made with the letter F, which we used to designate ours from those belonging to others.

A long rope with a noose and slip knot was fastened round the neck of the mule, the other round a tree; two men with another rope twisted it about its legs, when with a sudden jerk it was thrown to the ground; the red hot iron is now applied to the fleshy part of the hip—a terrible kicking and braying ensues, but it was always the sign that the work was done effectually.

In California, the most beautiful and valuable saddle horses are branded with a large unseemly mark on some prominent part of the body or neck, which would in this locality depreciate the value of the animals. I selected an Indian pony for myself; he was recommended as being a first rate buffalo horse; that is to say, he was trained to hunt buffaloes. This animal was given into my own charge, and I only then began to realize that I had entered into duties which I was unqualified to perform. I had never saddled a horse myself. My sedentary employment in a city, never having required me to do such offices; and now I was to become my own ostler, and ride him to water twice a day, besides running after him on the prairie for an hour sometimes before I could catch him. This onerous duty I finally performed as well as my companions. But, dear reader, follow me to a camp on the mountains of snow, where I exchanged my horse for a mule, at daylight, with the thermometer 20° below zero. Do you see, far away on the hill-side, an animal moving slowly? that is my mule; he is searching among the deep snows for a bite of blighted grass or the top of some wild bush to break his fast on. How will you get him? I will go for him; watch me while I tramp through the frozen snow. My mule sees me, and knowing that my errand is to prepare him for his day's journey, without first giving him provender to enable him to perform it, prefers to eat his scanty breakfast first, and moves leisurely along; his lariat, about thirty feet in length, trails along the

CATCHING A MULE.

ground. I have reached it, and at the moment I think I have him securely, he dashes away at a full gallop, pulling me after him through the snow; perfectly exhausted, I loose my hold; my hands lacerated and almost frozen. I lie breathless on the icy carpet. I am now a mile from camp, and out of sight of my companions. I renew my exertions, and gently approach him; this time he stands quiet, and I gather the rope in my hand, and pat him for a few minutes, and then mount him bare backed. The life and activity he possessed a few moments before, is entirely gone; he stands like a mule in the snow, determined not to budge a step. I coax, I kick him. I use the other end of the rope over his head; he dodges the blow; but his fore-feet are immovably planted in the snow, as if they grew there. I, worn out, and almost frozen, remain chewing the cud of bitter reflection, until one of my comrades comes to seek and assist me; he goes behind the mule and gives him a slight touch *à posteriori;* when, awakening from his trance, he starts at a hard trot into camp, quietly submits to be saddled, and looks as pleasantly at me as if he were inquiring how I liked the exercise of catching him. Similar scenes occurred daily; if it were not with myself it was with another. "Stubborn as a mule," is an o'er true adage, as I can fully testify.

A general examination of the equipage resulted in the knowledge that everything requisite for our journey, had been procured, and scales were in requisition to apportion the weight of luggage; 65 to 90lb. for each mule. The personal luggage of the men was restricted to a certain number of pounds—and all useless apparel, books etc., etc., were packed up and sent back to town.

We intended to pack on mules all the way, and it was necessary to take as little as possible of what we did not absolutely require.

A trial start was made, and the cavalcade started in excellent order and spirits, and we camped at the Methodist mission, about six miles from Westport.

CHAPTER IV.

Shawnee Mission—Indisposition of Col. Fremont—He returns to Westport—The Expedition proceeds onward—Sunrise on the Prairies—Meeting of our Delawares—Pottawatomie trading Post—Author turns Carpenter—Expediency—A Kansas Blacksmith—"Astonishment"—Quarter Master—Persecution—Success against Conspiracy.

We remained at the Methodist Mission until the next day, when we proceeded to the Shawnee Mission, a few miles further, and camped for the night. It was at this spot that Mr. Max Strobel made his appearance. He had been attached to Col. Stevens' expedition, but had left it on account of some misunderstanding with the officer in command. He requested Col. Fremont to allow him to accompany his expedition as a volunteer, and he would contribute his services as assistant topographer, &c. Col. Fremont hesitated, as his company was complete, but finally yielded to his continued entreaties. Col. Fremont, who had been slightly indisposed during the day, finding himself worse, decided to return to Westport, requesting us to continue on our journey until we met the Delawares, and then to encamp and await his return. The Col. returned to Westport, accompanied by Mr. Strobel, for whom it was necessary to purchase an outfit.

24th.—We travelled during this day on the open prairie. The weather was hazy and considerable rain fell during the last twenty-four hours; we camped on the open plains for the first time. At dawn of day I was up; I found the weather perfectly clear; and in breathless expectation of seeing the sun rise, I saddled my pony, determined to ride away from the camp—made my way through the long grass, for a considerable distance, before I perceived any inclination on the part of the majestic king of day to awake from his royal couch. Gradually the eastern horizon assumed a warmer hue, while some floating clouds along its edge, developed their form against the luminous heavens. The dark grey morning tints were superseded by hues of the most brilliant and gorgeous colors, which almost as imperceptibly softened, as the glorious orb of day commenced his diurnal course, and illumined the vault above; a slight rustling of the long grass, caused by a deliciously pleasant zephyr, which made it move in gentle undulation, was all that disturbed the mysterious silence that prevailed. I alighted from my pony, and gave him the range of his lariat. I perceived, that he preferred a breakfast of fresh grass, to the contemplation of the sublime scene around me, to which he seemed totally indifferent.

My heart beat with fervent anxiety, and whilst I felt happy, and free from the usual care and trouble, I still could not master the nervous debility which seized me while surveying the grand and majestic works of nature. Was it fear? no; it was the conviction of my own insignificance, in the midst of the stupendous creation; the undulating grass seemed to carry my thoughts on its rolling surface, into an impenetrable future;—glorious in inconceivable beauty, extended over me, the ethereal tent of heaven, my eye losing its power of distant vision, seemed to reach down only to the verdant sea before me.

There was no one living being present with whom I could share my admiration. Still life, unceasing eternal life, was everywhere around me. I was far away from the comforts of my home, not even in sight of a wigwam of the aboriginal inhabitants of the forests.

A deep sigh of longing for the society of man wrested itself from my breast. Shall I return, and not accomplish the object of my journey? No. I cannot; does not the grass, glittering in the morning dew in the unbroken rays of the sun, beckon me a pleasant welcome over its untrodden surface. I will onward, and trust to the Great Spirit, who lives in every tree and lonely flower, for my safe arrival at the dwelling of my fellowmen, far beyond the invisible mountains over which my path now lies.

27th.—To-day we met our Delawares, who were awaiting our arrival. A more noble set of Indians I never saw, the most of them six feet high, all mounted and armed *cap-a-pie*, under command of Captain Wolff, a "Big Indian," as he called himself. Most of them spoke English, and all understood it. "Washington," "Welluchas," "Solomon," "Moses," were the names of some of the principal chiefs. They became very much attached to Col. Fremont, and every one of them would have ventured his life for him.

Near the principal town of the Pottawatomies we remained encamped until the end of September, awaiting Col. Fremont. Two or three stores with no assortment of goods, and about thirty shanties make up the town. I went to every house in the place for a breakfast, but could not get anything to eat except some

Boston crackers, ten pounds of which (the whole supply in the town) I bought. My ride into the town was for the purpose of having strong boxes made to carry my daguerreotype apparatus. The baskets in which they had been packed being broken and unfit for use. There was not a carpenter, nor any tools to be had in town. There was a blacksmith's about ten miles from town, where it was likely I could procure them. It being absolutely necessary that I should have the boxes, I induced one of our Delawares to accompany me, carrying on our horses a sufficient quantity of dry goods box covers and sides to manufacture them. When we arrived at the blacksmith's house, the proprietor was absent. His wife, an amiable woman, prepared dinner for us, and gave us the run of the workshop, where I found a saw and hatchet; with these instruments I made the boxes myself, and by the time they were finished, the blacksmith returned. He refused to receive pay for my dinner, but charged for the nails, raw hide, etc., I covered the boxes with, and the use of his tools. The lady told me I was the first white man she had seen, except her husband, in three years. I gave some silver to the children, and mounting our horses, with a huge box before us on our saddles, we slowly retraced our way to camp, where we arrived at dark. Nobody in camp knew my errand to town, and I never shall forget the deep mortification and astonishment of our muleteers when they saw my boxes. All their bright hopes that the apparatus would have to be left, were suddenly dissipated. The expenses attendant on the manufacture of the boxes, and the material, were nearly five dollars, which I requested our quarter-master to pay, as Col. Fremont left him money for disbursements;

he refused, at first, but was finally induced to do so under protest. I have every reason to believe that my baskets were purposely destroyed; and but for my watchful and unceasing care, they would have been rendered useless. The packing of the apparatus was attended with considerable trouble to the muleteers, and also to the officer whose duty it was to superintend the loading and unloading of the mules; and they all wanted to be rid of the labor. Hence the persecution to which I was subjected on this account. Complaints were continually being made to Col. Fremont, during the journey, that the weights of the boxes were not equalized. Twice I picked up on the road the tin case containing my buff, &c., which had slipped off the mules, from careless packing—done purposely; for if they had not been fortunately found by me, the rest of the apparatus would have been useless. On one occasion, the keg containing alcohol was missing; Col. Fremont sent back after it, and it was found half emptied on the road.

I am induced to make these remarks to show the perseverance and watchfulness I had to exercise to prevent the destruction of the apparatus by our own men.

CHAPTER V.

Letter to W. H. Palmer—Col. Fremont's Return to St. Louis, and increased Illness—Expedition under charge of Delawares—Camp Proceeds to "Smoky Hills"—Fort Riley—Solomon's Fork—First Buffalo—Barometers go on a Buffalo Hunt—Encampment on "Salt Creek"—Indian Method of Cooking Buffalo Meat—Olla Podrida—Wasting of Provisions—Kinnikinick—Havana Segars—Indian Amusements—Camp Life—Hewers of Wood and Drawers of Water—Author's Opinion of Col. Fremont—He Nominates him for the Presidency.

After remaining at this camp two days, Mr. Strobel arrived with a letter from Col. Fremont to Mr. Palmer, stating that his increasing illness made it necessary that he should return to St. Louis for medical advice, and directed us to proceed as far as Smoky Hills, and encamp on the Saline fork of the Kansas River, where there were plenty of buffalo, and remain there until he joined us, which he hoped would be in a fortnight.

The expedition, during encampment, was to be under the supervision of Mr. Palmer. Accordingly, we continued our journey, and crossed the Kansas River at its junction with the Republican, within half a mile from Fort Riley, thence to Solomon's Fork, in crossing which creek, some of the baggage of the camp became saturated with water.

Immediately after crossing Solomon's Fork, we saw our first buffalo. As soon as he was discovered, our Delawares gave a whoop, and they all started, helter skelter, the officers and muleteers following, leaving the baggage animals to take care of themselves. Our engineer, Mr. Egloffstien, after the first excitement had passed, suddenly drew rein—I did so likewise.

He remarked, "I have been at full speed for a mile, with both barometers slung across my back."

I never saw any one look so alarmed as he did. I had exchanged ponies, to give him an easy-going animal, so as not to shake the instruments, and now his rashness had probably injured them. He alighted and examined them; luckily, they were well packed with cotton, and they were not at all disarranged. Our buffalo was soon killed; and that night we made an encampment on a beautiful site near Salt Creek, and about four miles from the Kansas River, with buffalo steaks for supper.

[Extract from a Letter.]

Dear S——:

We are now encamped, as it were, for a pleasure excursion, for all the day is employed in hunting, gunning, shooting at a mark with rifles, and preparing buffalo meat in all the modes in which it is said to be good.

I was much amused, the first day we encamped here, to see the Indians go into the woods on the creek, and bring out straight green sticks, the size of a small walking-cane, and proceed to divest them of their outer peeling—also pointing them at both ends.

I soon discovered their use: they cut the buffalo meat in strips about an inch thick, four wide, and twelve to fifteen long. The stick is then inserted in the meat, as boys do a kite stick; one end of the stick is then stuck in the ground, near the fire, and the process of roasting is complete—the natural juice of the meat is retained, in this manner, and I think it the most preferable way to cook game. The breast of a fat antelope prepared thus is a most fitting dish for a hungry man.

Several kinds of game were brought into camp this evening, buffalo, antelope, and deer, by the Indians, and our most successful gunner, Mr. Fuller, brought in two wild turkeys, three ducks, a rabbit, and a prairie hen, the result of his day's sport. Our cook for the nonce is making a splendid Olla Podrida. This is our first week in camp, and we are living sumptuously— coffee, tea, and sugar three or four times a day.

I have no control of the commissariat department, but I very much fear that we shall want some of the good things which are now being inconsiderately wasted. Our quarter-master is determined to enjoy himself—his motto is "dum vivimus vivamus."

While I am writing, I am smoking a pipe filled with "Kinnikinick," the dried leaves of the red sumach; it is pleasant and not intoxicating—a very good substitute for tobacco. The Delawares have been preparing some for their journey. They smoke it mixed with tobacco.

My quarter-box of Havanas are all gone, already; they were the only ones in camp, and every time I took out my pouch, I of course handed it round to my companions, which soon diminished my store. I close this letter by giving you a description of an Indian game, which our Delawares participated in last night.

A large fire of dried wood is brightly burning—around it sit, cross-legged, all our Delawares; behind them are the rest of us, standing looking on. I contributed the article (which was a large imitation seal ring, several of which I bought to exchange with the Indians for moccasins) with which they amused themselves. One of them took the ring, and while the rest are chanting Highya, Highya, he makes sundry contortions of his limbs, and pretends to place it in the hands of the one next to him. This one goes through with the same antics, until all have had it or are supposed to have had it. The first one then guesses who has the ring; if successful, he wins the ring; if not, he contributes tobacco for a smoke; a pipe is filled, which is generally a tomahawk with a bowl at the butt-end; the handle is hollow, and communicates with the bowl, thus forming a weapon of war, as well as the calumet of peace; each one takes two or three puffs and then passes it around.

Dear S——:

The duties of camp life are becoming more onerous as the weather gets colder. It is expected that each man in camp will bring in a certain quantity of fire-wood! My turn came to day, and I am afraid I shall make a poor hand in using the axe; first, I have not the physical strength, and secondly, I do not know how. I managed by hunting through the woods to find several decayed limbs, which I brought in on my shoulder. I made three trips, and I have at all events supplied the camp with kindling wood for the night.

I certainly, being a "Republican," do not expect to warm myself at the expense of another; therefore, arduous as it is, I must, to carry out the principle of equality, do as the rest do, although it is not a very congenial occupation.

* * * * * *

'Tis very strange how fallacious ideas of mankind obtain stronghold in the minds of those who should know better. The night previous to leaving home, I was asked how I could venture my life with such a man as Col. Fremont? "A mountaineer"—"an adventurer"—"a man of no education."

During my voyage up the Missouri, I had continual opportunities of conversing with Col. Fremont.

If you ever see Mr. —— and Mrs. ——, please say to them, that the character of Col. Fremont as a gentleman of "high literary attainments," "great mental capacity," and "solid scientific knowledge," is firmly established in my own mind.

These personal observations, added to the knowledge I gained of him from report, has brought me to the conclusion that he is not only a "man of education," but a "man of genius and a gentleman." One would suppose that the "conqueror of California," "the successful commander and governor," would have a little to say about himself—some deeds to vaunt of—some battle to describe. I found him reserved, almost to taciturnity, yet perfectly amiable withal. No one, to see him, would ever imagine that a man of great deeds was before him.

My estimation of character is seldom wrong. I may have been imprudent in undertaking this journey, which already "thunders in the index," and on which I shall have to encounter many personal difficulties; but, if I felt safe enough to impulsively decide to accompany him, without personally knowing him—how much safer do I now feel from the short time I have known him!

All the men in camp have the same opinion of him.

Yesterday, while discussing the merits of the most prominent men who were likely to be placed before the people for the "next President," I mentioned the name of "Col. Fremont." It was received with acclamation, and he is the first choice of every man in camp. So you see I am safe enough with the man—it is only the mountains which are the "stumbling blocks." Yet I have full faith that I shall return once more to you in safety.

*   *   *   *   *   *

CHAPTER VI.

Kansas Territory—Arkansas and Kansas Rivers—Tributaries—Timber—River Bottoms—Varieties of Game—Cereals—Coal—Geological Formation—Adventure in the Woods—Wild Grapes—Indian Method of procuring them—Brandy vs. Poison—Return of the Author's Brandy-flask—He turns Washerwoman—Novel Mode of Mangling Clothes—Lost Mule—Beaver Trappers—Rifle Practice.

KANSAS lies between the thirty-seventh and fortieth degrees of north latitude. The Indian Territory bounds it on the south, Utah and New Mexico on the west, Nebraska on the north, and Missouri on the east.

There are numberless streams of water in the Territory. The Arkansas which rises in the Rocky Mountains, runs nearly six hundred miles through it. Kansas River, which empties into the Missouri near Kansas City, has many forks of considerable size, viz., the Republican, Solomon Fork, Grand Saline Fork, Vermilion, Little Vermilion, Soldier Creek, Grasshopper Creek, Big Blue, Pawnee Fork, Walnut Creek, Wakarusa, and several others. The country is well watered, and on all the rivers grows timber of large size and in great variety. The river bottoms are very fertile, being covered with an alluvial black soil from twelve to twenty-four inches deep. These bottoms vary in width from four to seven miles.

Another bottom over which the waters must have once flowed, is elevated about sixteen feet from the river, and high up some sixty to seventy feet, lies the immense

PRODUCTIONS OF THE COUNTRY.

undulating prairie, teeming with buffalo, blacktail, deer, antelope, sage, and prairie chickens. Thousands of cayottes—a small wolf, make night hideous with their shrill discordant bark. The large white wolf is also found in great numbers on the rivers. We killed wild turkeys and ducks. The second bottoms are studded with groves of timber. The various kinds of oak, maple, elm, red-flowered maple, black walnut, locust, beech, box, elder, wild-cherry, and cotton-wood, attain a large size, and are to be found on the Kansas River and its many tributaries in quantities.

Grasses of a hundred different kinds, some of them rank and high, but the most of them possessing highly nutritive qualities, grow spontaneously on the prairies, and afford nourishment to immense quantities of game.

The water of the Kansas partakes in color of the character of the soil over which it passes. It is, I am inclined to believe, always turbid. I found it quite unfit for' daguerreotype purposes, and had to preserve many of my plates until we approached the crystal streams from the Rocky Mountains, to finish them. During our long camp on Salt Creek, our topographical engineer and myself explored the country for miles. Coal in abundance is to be obtained with but little exertion; in many instances it crops out on the surface of the ground. The general character of the formation of this country is the same as Missouri—a secondary limestone.

Dear S——:

To-day we had a delightful jaunt through the woods which fringe the forests of Salt Creek. Cotton-wood, oak, elm, ash, hickory, grow luxuriously, some of them to an immense height. Our Delaware that accompanied Egloffstien and myself suddenly stopped, and pointed upward. There, at a height of over one hundred feet, suspended between two oaks, were grape-vines loaded with rich luscious looking fruit.

How were we to obtain them? I could not climb so tall a tree. Mr. Egloffstien declined, and we both depended on our Delaware. He looked very grave and said: "Suppose Delaware want grapes, he know how to get them."

By this time our desire increased to obtain the prize, which seemed to say, "Come and take me." I commenced climbing one tree, and my friend the other. When we had exerted ourselves, and had reached the first limb, on which we stopped to rest, we heard a grunt from our Delaware, and almost at the same moment, the whole vine came tumbling down on his head.

He purposely waited until we were in the trees, to see how "white men gathered grapes." He took hold of the grape vine, and with one tremendous pull, down it came; when we descended, he was quietly stowing away the choicest bunches in his hunting shirt. I never would have dreamed of destroying such a noble vine, to gratify my appetite.

The grapes were small, but sweet and well flavored. I ate a great many of them. I had been without fruit or vegetables for four weeks, and they were very grateful to me. I hope I shall not suffer for my imprudence. Good night.

BRANDY *versus* POISON.

Previous to leaving New York, I had two tin flasks made, to contain about a quart each, which I intended to have filled with alcohol for daguerreotype purposes. At Westport, I purchased a quart of the best quality of old cognac, filled one of them for medicinal purposes, and carefully packed my flask in my daguerreotype boxes. One day during our camp at Salt Creek, one of our Indians being ill, I opened my flask and pouring out about an ounce, replaced it. I noticed, however, that a chemical action had taken place, turning the brandy exactly the color of ink. One of our mess saw me open my box and appropriate a portion of the contents of the bottle; I am not certain but that I tasted it myself.

The next day I had occasion to go to my box, when to my utter astonishment, my flask of brandy was gone. I immediately suspected the very person who afterwards proved to be the thief. Keeping my loss a secret, at dinner I carefully watched the countenances and actions of the whole party, and the effects of liquor were plainly visible on the person of this man.

"How excellent," said I, "would a bottle of old cognac be as a digester to our tough old buffalo bull.— Gentlemen, how would you like a drop?" "Bring it forward by all means, Carvalho. You have, I verily believe, Pandora's box; for you can produce everything and anything at a moment's notice, from a choice Havana to old brandy."

"With your leave, gentlemen, I will procure it. I have two flasks exactly alike; one contains poison, a mixture of alcohol, and some poisonous chemicals for making daguerreotypes; the other contains the best brandy to be had on the Kansas River."

I went to my box, and turning up my hands with an exclamation of surprise, announced to the mess that the "bottle containing the poison, and which I laid on the top of my box last night, is missing." Like Hamlet I looked into the face of the delinquent, and I never shall forget his expression when I remarked that "the liquid in the purloined flask was poison, and perfectly black, and although it would not kill immediately, an ounce will produce certain death in 48 hours."

"Gentlemen! I shall, in consequence, have to reserve the brandy to make another similar mixture, to substitute for alcohol; therefore I am sorry I cannot treat you as I intended."

Of course the innocent parties felt indignant that my flask had been stolen, and that one of their party was suspected.

The thief was discovered, although he nor any one else knew that I detected him. The next day I went to my box again, and in its proper place, I found my brandy flask about half full. Our friend had taken several strong pulls during the night and morning, and likely enough he looked at the contents, and finding them black as ink, believed all about the poison, and fearing to die, replaced the flask, without detection. When I discovered it, I showed it around and also the color of the contents, and told them it was not poison but "good old brandy." I tasted a little, and divided it among the party.

The man that took it knew I suspected him, and his whole conduct to me during the journey, was influenced by that event, although I never taxed him with it.

DEAR S——:

Yesterday being a fine, mild day, I thought I would examine my wardrobe, and have such articles as I had worn during the last three weeks washed. I col lected three shirts, as many pairs of stockings, together with handkerchiefs and drawers; I made up a dozen pieces; and I assure you, that how or by whom they were to be washed, never entered into my mind. I offered some compensation to one of our muleteers if he would wash them, but he was perfectly independent of the necessity of earning money in that way. I soon discovered, that I would have to become my own washerwoman; and obtaining some soap from the quarter master, I gathered up my duds, and made my way down the banks of the creek, to a convenient place, and there I entered upon my novitiate. I rubbed the skin off my hands during the operation, but after considerable application, I succeeded in cleansing them, and hung them out to dry. I doubled them up, and laid them carefully under my buffalo robe couch, last night; and this morning they are as smooth as if they had been "mangled." To-day I employed myself making a pair of buckskin mitts and moccasins, as I shall require them before many weeks; most of the Indians and muleteers are out, looking for a large black mule, the finest animal in the collection, which was missing last night.

Yesterday two beaver trappers came into the Delaware camp, and traded for sugar and coffee with the Delawares. I have my suspicions that our mule conveyed them away, as they are no longer on the creek where they set their traps yesterday.

I must leave off my journal, as it is my usual hour for rifle practice; I have become quite an expert; at one hundred paces, I have hit the "bull's eye" twice in five times, which is not bad shooting, considering I have had no practice since I was a member of a rifle volunteer company in Charleston, some twenty years ago.

CHAPTER VII.

Delaware Medicine Man—Illness of Capt. Wolff—Author turns Doctor—Empty Commissariat—Expedition to Fort Riley for Fresh Supplies—Professor Espy's Theory of Rain—Indians on Kansas Prairies—Sleet and Snow Storm—Tent Blown Down—Approach of Cold Weather—"Pony Missing."

DELAWARE MEDICINE MAN.

For several days, Capt. Wolff, the chief of our Delawares, had been ailing, this morning I noticed some unusual preparations in their camp, on inquiring I was told that, in the woods, Capt. Wolff, who was very sick, was undergoing the Indian ceremony of "incantation," by one of the tribe, who was "a great medicine man." The ceremony was conducted in secret, but I found out afterwards the place, and from the mode which was explained to me, I understood the rite perfectly. A small lodge, composed of the branches of trees, high enough for a man to sit upright in, was built; in this the patient was placed in a state of perfect nudity. "The Medicine Man," who is outside, takes a "pipe," filled with "kinnikinick and tobacco," and hands it in to the patient. While the Medicine Man recites the "all powerful words," the patient puffs away until the lodge is filled with smoke; when the poor devil is almost suffocated, and exhausted, he is taken out, wrapped in his blankets, and conveyed to his own lodge.

Feeling anxious about him, I went in to see him

about an hour afterwards; I found him in a high state of febrile excitement, which had, no doubt, been increased by his extraordinary treatment; he complained of dreadful headache and pain in his back. He thought he was going to die. I told him if he would submit to my advice I thought I could cure him—he consented, and I administered ten grains of calomel, and four hours afterwards half oz. of Epsom salts. He is now considerably relieved; and I think by the morning he will be well. Indigestion was the cause of his suffering. I made him some of the arrowroot, which thanks to your usual foresight I found stowed away in my trunk. I shall reserve it for similar occasions.

Col. Fremont has not yet arrived.

Our quarter master has suddenly discovered that his commissariat is empty, and talks of sending to Fort Riley for fresh supplies to-morrow; if he does I will forward a package of letters to you, which please preserve from public eye.

Two Delawares and a muleteer are now preparing to go to "Fort Riley" for supplies. Capt. Wolff is better; by evening I hope he will be perfectly well. I think if I had not treated him he would have probably died. Another "incantation" would certainly have killed him. I shall continue to write to you. Most probably we shall be detained here a week longer; it is now the 20th October, and I am afraid Col. Fremont is seriously ill; you will, of course, have heard of his return, and I shall look forward to receive by him happy tidings from all those I love.

PROFESSOR ESPY'S THEORY OF RAIN.

I have had occasion to observe that the immense clouds of smoke which filled the atmosphere continually during the time the prairies were on fire, were condensed during the cold of the night, sometimes forming rain, but always heavy dew, which I did not observe before the prairies were burning.

I think Prof. Espy says that artificial rain can be produced by smoke from large fires, and from the observations I have made I coincide in that theory.

It is not unlikely that the Indians, who have from the earliest knowledge of the prairie country annually set the high rank grass on fire, did it to afford artificial moisture for the immense tracts of buffalo grass plains, on which subsist hundreds of thousands of buffalo, elk, and deer. No rain falls at certain seasons, and without dew the grass would be all burnt up by the scorching heat of the sun.

The Indians, I believe, practically put into operation the theory of Espy—knowing from experience that smoke is condensed into dew.

On the Kansas River the dew fell very heavily. I found it necessary while doing guard to cover myself with my India-rubber poncho, to prevent my clothes from becoming saturated with water.

*   *   *   *   *   *   *

Last night our camp was visited with a heavy storm of rain and sleet; it was bitter cold. It rained considerably yesterday, but the temperature was not lower than 65°. The wind increased during the night, and one sudden gust blew our cotton tent completely over, exposing us to the peltings of the merciless storm of sleet. Several of us essayed to raise the tent, but the ground had become saturated with moisture, and afforded no hold for our tent-pins, and we consequently lay down, wrapped ourselves in our India-rubber blankets, and bewailed our fate.

We presented an interesting picture when the daylight came. Many of our clothes, which were lying loosely in the tent, were blown some distance from camp, and we were all drenched to the skin. The weather cleared off at sunrise, and around a large camp-fire we dried our clothes and passed jokes on each other's distressing appearance. Winter seems to have suddenly set in; the thermometer indicated, at sunrise, 34° ("*por peccados,*" as the Spaniards say.) Many of our animals pulled up their picket-pins, and sought shelter in the woods. My pony is missing, among others, and on *myself and on no one else* devolves the delightful duty of finding him. I have put on, for the first time, my waterproof boots, as I have a wet road, and, probably, a long distance to walk, before I find my horse. He is safe enough on the creek; the Indians saw him while hunting up theirs.

CHAPTER VIII.

Author's First Buffalo Hunt—Pursuit—Perilous Situation—Mode of Attack by the Indians—Solitary and Alone—Pony killed for Food—An Ill Wind that blows Nobody any Good—Incredulity of Indian Hunters—Return to Camp—Prairies on Fire—Suffocating Smoke—Mr. Egloffstein on a Wolf Hunt—Fire Light and Moonlight—Camp surrounded by Fire—Dangerous Situation—Arrival of Colonel Fremont—Preparations to resume our Journey—Escape through the Blazing Element.

MY FIRST BUFFALO HUNT.

AT daylight, on the 25th October, the hunters were at breakfast. At our mess, feats of daring and gallant horsemanship were being related, while our roast was preparing. Weluchas, a most successful hunter, and as brave and daring an Indian as ever fashioned a moccasin or fired a rifle, approached me, remarking, "What for you no hunt buffalo—got buffalo pungo?" (horse.) I had, while at breakfast, almost made up my mind to go—this, however, determined me. In quick time I had my horse saddled, and, fully equipped with rifle, navy revolver, and sheath knife, was all ready for a start. On this occasion our party consisted of eight Delawares and four white men, besides myself. I rode out of camp, side by side with Weluchas, who seemed gratified that I accompanied him. The buffaloes, from having been daily hunted for several weeks, had gone South about fifteen to twenty miles, and we had to ride that distance before we saw any game.

A BUFFALO HUNT. 51

DISCOVERY OF A HERD.

After about three hours' gentle trotting, one of the party started a cayote, and we chased him until he disappeared in the brush. When we reached the brow of a hill, Weluchas ejaculated, in deep, low tones, "Buffalo," "big herd"—"plenty cow." I turned my eyes, and, for the first time, beheld a large herd of buffaloes occupying an extensive valley, well wooded and watered, and luxuriant with the peculiar short curled grass, called "Buffalo grass" (*Lysteria Dyclotoides*), on which this animal principally feeds. I gazed with delight and astonishment at the novel sight which presented itself. There must have been at least 6,000 buffaloes, including cows and calves. It was a sight well worth travelling a thousand miles to see. Some were grazing, others playfully gambolling, while the largest number were quietly reclining or sleeping on their verdant carpet, little dreaming of the danger which surrounded them, or of the murderous visitors who were about to disturb their sweet repose.

THE FOE DISCOVERED BY THE SENTINEL—THE HERD IN MOTION.

Taking the word of command from Capt. Wolff, one of the finest proportioned men I ever beheld, we kept silent, to await the direction which the herd would take when they discovered us. An old bull was stationed several hundred yards in advance of the herd, as sentinel: they invariably follow him, as leader, even into danger. He soon espied us; and suddenly, as if by magic, the whole herd was in motion. We occupied such a position that they passed within rifle shot.

THE PURSUIT.

At a signal, the whole party, except myself, galloped after them. I was intensely absorbed by this mighty cavalcade passing with majestic stride, as it were, in review before me. My pony, anxious for the chase, fretted and champed at the bit. I singled out what I thought a fat cow (the bulls are tough and hard, and are only hunted by the Indians for their robes—their flesh never being eaten when cows can be obtained), and in a few seconds, I was riding at full speed. It requires a very fleet horse to overtake a buffalo cow. A bull does not run quite so fast. After a chase of about two miles, I was near enough to take sight with my rifle, by stopping my pony. I fired and wounded him in the leg—reloaded, and started again at full speed, the buffalo running less swiftly. I fired again, but this time without effect. Not wishing him to get too far ahead of me, I took out my revolver, and got within pistol shot, when I discovered I had chased an old bull instead of a cow.

PERILOUS SITUATION.

I fired my pistol four times at full speed, and was endeavoring to sight him again, when the bull suddenly turned upon me, within five yards of my horse. My well-trained pony instantly jumped aside. The bull, in turning, got his wounded leg in a painful position, and stopped, which gave me time and opportunity to save my life; for, with my total inexperience, I should not have been able to have mastered him. My horse jumped aside without any guiding from me, having been trained to this by the Indian from whom we purchased him. I reloaded my rifle, and took deliberate aim at a vital part. When dying, I approached the monster that had given me such a fright, when he turned his large black eyes mournfully upon me, as if upbraiding me with having wantonly and uselessly shot him down.

MODE OF ATTACK BY THE DELAWARES.

A Delaware Indian, in hunting buffaloes, when near enough to shoot, rests his rifle on his saddle, balances himself in his stirrup on one leg; the other is thrown over the rifle to steady it. He then leans on one side, until his eye is on a level with the object, takes a quick sight, and fires while riding at full speed, rarely missing his mark, and seldom chasing one animal further than a mile.

SOLITARY AND ALONE.

After recovering from my fright, and the intense excitement incidental to the chase, other sensations of a different character, although not less disagreeable, immediately filled my mind. I discovered that I was entirely alone, in an uninhabited, wild country, with not a human being in sight. I had chased my bull at least five miles. My companions had taken a different direction, nor was a single buffalo to be seen. My mind was fully alive to the perils of my situation. I had left my pocket compass in camp, and I did not know in what direction to look for it. I mounted my horse and walked to the top of a hill to see if I could find any traces of the party. I discovered looming in the distance, Smoky Hills some twenty miles off. My mind was in a slight degree relieved, although I was almost as ignorant of my geographical position as I was before. I did not despair, but unsaddling my horse, I gave him an hour's rest; the grass was fresh, and he appeared totally unconcerned at my situation.

PONY KILLED FOR MEAT.

Poor fellow! Little did I think that day, as he carried me, so full of life and high spirit, that in a few weeks he would be reduced to a mere skeleton, and that I should be obliged, in order to save my own life on the mountains of snow, to partake of his flesh. I shed tears when they shot him down, and I never think of his generous, willing qualities, but I lament the stern necessity that left his bones bleaching on the mountains.

IT'S AN ILL WIND THAT BLOWS NOBODY GOOD.

I re-saddled my pony, and turned his head in the direction of Smoky Hills, fervently hoping to fall in with some of our party; nor was I disappointed, for after riding about an hour, I discovered to the left of my course a horse without a rider. As I approached it, I recognized the animal, and in a little while I saw its owner, my friend Weluchas, walking slowly, with his eyes intently fixed on the ground. He told me he was looking for his tomahawk pipe, which he had dropped while hunting. I joyfully assisted him in finding it, after a persevering search of an hour. He had been at least an hour on the spot before I came up. To this lucky circumstance I attributed my arrival in camp that night, for when we resumed our journey, he took a course some six points variation from the one I was travelling. On our way we fell in with Capt. Wolff and another Delaware, who were busily engaged cutting up a fine fat cow. I was soon at work, but I gave up after an ineffectual attempt to cut the liver, which is very delicate eating, my knowledge of human anatomy not being of any service to me in dissecting buffaloes.

THE INCREDULITY OF THE INDIAN HUNTERS.

While journeying campwards I related to the party my adventure with the old bull. I, of course, finished it by stating I had slain him. Capt. Wolff looked at me with a most quizzical and incredulous smile, and emphatically remarked, in his broken English, "Carvalho no kill buffalo." I insisted that I had left him dead on the field. At this the whole party laughed at me. I felt annoyed, but soon found it was no use to contend with them. Weluchas, who was really my friend, and to whom I had rendered several services, such as bleeding him and curing him of fever, could not believe the statement I had made. Capt. Wolff, seeing me look offended, said, in these exact words:—"When Capt. Wolff kill buffalo, he cut out the tongue. Indian shoot buffalo, bring home tongue. Carvalho no bring buffalo tongue; he no kill buffalo." This was powerful argument, and the inference perfectly logical; and I soon changed the subject. Gentle reader, do you think I was equal to cutting out, by the roots, a tongue from the head of an old buffalo bull, after telling you that I did not succeed in getting out the liver of a young cow, after the animal was opened? Surely I was not; but even if I had been, the alarming situation I found myself in, at the time he fell, prevented me from attempting it, if I even had known it was the hunters' rule to do so.

RETURN TO CAMP.

My messmates, to whom I related my adventure, had not the slightest idea that I had lost my way in the chase. I came into camp with the rest of the party, that night, about seven o'clock, tired and hungry. After eating a hearty supper, I wrapped myself up in my blankets and was soon asleep, dreaming of the disputed honors I had gathered in my maiden hunt after a buffalo bull.

PRAIRIE ON FIRE.

Oct. 30.—During the day, the sun was completely obscured by low, dark clouds; a most disagreeable and suffocating smoke filled the atmosphere.

We were still encamped on the Saline fork of the Kansas River, impatiently awaiting the arrival of Col. Fremont, who had not yet returned from St. Louis. His continued absence alarmed us for his safety, and the circumstance that the prairies were on fire for several days past, in the direction through which he had to pass to reach us, added to our anxiety.

Night came on, and the dark clouds which overhung us like an immense pall, now assumed a horrible, lurid glare, all along the horizon. As far as the eye could reach, a belt of fire was visible. We were on the prairie, between Kansas River on one side, Solomon's Fork on another, Salt Creek on the third, and a large belt of woods about four miles from camp on the fourth. We were thus completely hemmed in, and comparatively secure from danger.

Our animals had been grazing near this belt of woods, the day before, and when they were driven into camp at night, one of the mules was missing. At daylight a number of our Delawares, Mr. Egloffstein, our topographical engineer, and myself, sallied out in search of it.

After looking through the woods for an hour, we discovered our mule lying dead, with his lariat drawn close around his neck. It had become loose, and, trailing along the ground, got entangled with the branches of an old tree, where in his endeavors to extricate himself he was strangled.

We were attracted to the spot by the howling of wolves, and we found that he had been partially devoured by them. Our engineer, who wanted a wolfskin for a saddle-cloth, determined to remain to kill one of them.

I assisted him to ascend a high tree immediately over the body of the mule, untied the lariat, and attaching his rifle to one end of it, pulled it up to him.

The rest of the party returned to camp. About four o'clock in the afternoon, he being still out, I roasted some buffalo meat and went to seek him. I found him still on the tree, quietly awaiting an opportunity to kill his wolf.

A heroic example of perseverance on an eminence smiling at disappointment.

Mr. Egloffstein declined to come down; I told him of the dangers to which he was exposed, and entreated him to return to camp. Finding him determined to remain, I sent him up his supper, and returned to camp, expecting him to return at sundown.

About this time the prairie was on fire just beyond the belt of woods through which Col. Fremont had to pass.

Becoming alarmed for Mr. Egloffstein, several of us went to bring him in. We found him half-way to camp, dragging by the lariat the dead body of an immense wolf which he had shot. We assisted him on with his booty as well as we could.

My "guard" came on at two o'clock. I laid down to take a three hours' rest. When I went on "duty," the scene that presented itself was sublime. A breeze had sprung up, which dissipated the smoke to windward. The full moon was shining brightly, and the piles of clouds which surrounded her, presented magnificent studies of "light and shadow," which "Claude Lorraine" so loved to paint.

The fire had reached the belt of woods, and seemingly had burnt over the tree our friend had been seated in all day.

The fire on the north side had burned up to the water's edge, and had there stopped.

The whole horizon now seemed bounded by fire.

Our Delawares by this time had picketed all the animals near the creek we were encamped on, and had safely carried the baggage of the camp down the banks near the water. When day dawned, the magnificent woods which had sheltered our animals, appeared a forest of black scathed trunks.

The fire gradually increased, yet we dared not change

our ground; first, because we saw no point where there was not more danger, and, secondly, if we moved away, "Solomon," the Indian chief, who after conducting us to the camp ground we now occupied, had returned to guide Col. Fremont, would not know exactly where to find us again.

We thus continued gazing appalled at the devouring element which threatened to overwhelm us.

After breakfast, one of our Delawares gave a loud whoop, and pointing to the open space beyond, in the direction of Solomon's Fork, where to our great joy, we saw Col. Fremont on horseback, followed by "an immense man," on "an immense mule," (who afterwards proved to be our good and kind-hearted Doctor Ober;) Col. Fremont's "cook," and the Indian "Solomon," galloping through the blazing element in the direction of our camp.

Instantly and impulsively, we all discharged our rifles in a volley.

Our tents were not struck, yet we wanted to make a signal for their guidance. We all reloaded, and when they were very near, we fired a a salute.

Our men and Indians immediately surrounded Col. Fremont making kind inquiries after his health.

No father who had been absent from his children, could have been received with more enthusiasm and more real joy.

To reach us he had to travel over many miles of country which had been on fire. The Indian trail which led to our camp from "Solomon's Fork," had become obliterated, rendering it difficult and arduous to follow; but the keen sense of the Indian directed him under all difficulties directly to the spot where he had left us.

During the balance of the day, the camp was put in travelling order.

With the arrival of Col. Fremont, our commissariat had received considerable additions of provisions, more, in fact, than he had any good reason to suppose we had consumed during his absence.

The reverse was exactly the truth. The provisions intended for our journey had been lavishly expended, and surreptitiously purloined.

Twice it became necessary to send to Fort Riley to procure supplies.

The season had advanced, and it became imperatively necessary to continue onwards—we should have plenty of game until we got to Bent's Fort, where there always were kept large supplies of provisions, and where Col. Fremont intended to refit and replenish.

At midnight, the fire crossed the Kansas River. I was in a great state of excitement. I mounted my horse and rode out in the direction of the Kansas, to see if the fire had actually crossed; I suppose I must have advanced within half a mile, before I discovered that the prairie was on fire on this side of it. I turned round, and galloped as I thought, in the direction of camp, but I could not descry it. I continued onwards; but as there were woods all around Salt Creek, I had lost my landmarks, and was in a terrible quandary. I however reached Salt Creek, and with great difficulty returned to camp, after an absence of three hours.

At daylight, our animals were all packed, the camp raised, and all the men in their saddles. Our only escape was through the blazing grass; we dashed into it, Col. Fremont at the head, his officers following, while the rest of the party were driving up the baggage animals. The distance we rode through the fire, could not have been more than one hundred feet, the grass which quickly ignites, as quickly consumes, leaving only black ashes in the rear.

We passed through the fiery ordeal unscathed; made that day over fifteen miles, and camped for the night on the dry bed of a creek, beyond the reach of the devouring element.

CHAPTER IX.

Walnut Creek—Necessity of strict discipline—Neglect of duty—Horses stolen—Cheyenne Indians—Thieves overtaken—Watchfulness of Col. Fremont—Immense herds of Buffalo—Buffalo hunt on a large scale—Buffalo chips—Prairie dogs, Owls, &c.—Indians in camp—Raw Antelope liver.

The cold was intense during our last encampment at Walnut Creek. About an hour after the midnight watch had been relieved, and while the last watch were warming their benumbed limbs before a large fire, one of the men on horse guard left his duty, and came into camp to warm himself—Col. Fremont, who was always on the " qui vive," suddenly appeared at the camp-fire. This was not unusual, that he should personally inspect the guard, but he took such times, when he was least expected—in order to see if the men did their duty properly.

The Colonel accosted the officer of the watch, and enquired if Mr. —— had been relieved? He replied that he had not, but gave as an excuse, the coldness of the weather. Col. Fremont lectured the officer, and had another man immediately sent out to take his place. He was highly displeased, and as a punishment, told Mr. —— that he expected he " would walk," during the next day's travel. I had been relieved a short time before, and I knew how cold I was, and that it was necessary to move about continually, to keep up the

NECESSITY FOR STRICT DISCIPLINE. 63

circulation of the blood; under the circumstances, I thought the punishment disproportionate to the offence.

I was a novice in camp life among Indians, and was not aware of the stern necessity required for a strict guardianship of the animals; but the sequel proved, that the "slight dereliction" from duty, as I thought it, involved the most serious consequences.

At day-light, when the animals were driven in to be loaded and packed for the day's journey, five of them were missing. The camp was, in consequence, delayed, while the animals were sought; half the day was lost in an ineffectual search. Our Delawares reported having discovered moccasin prints on the snow, and at once decided they were made by Cheyenne Indians, from their peculiar form.

The next day we followed a track made by "shod horses," which convinced us we were on the right scent. The Indians do not shoe their horses.

On the "divide," near the Arkansas River, we saw one of our mules grazing, but so worn out by the hard drive, that he was unable to continue, and the Indians left him on the prairie.

It took us several days to reach the village, which was situated on the part of the Arkansas River known as Big Timber, near Mr. Bent's house.

At this village we found the rest of the animals, and some of the thieves. On examining them, they confessed that they had watched our camp during the night, for an opportunity to run off our animals, but found them guarded, until one man left his watch, and went to warm himself at the camp fire, during which time they stole five of them, and if they had had an hour longer time, they would have stolen a great many more. They went so far as to point out the very man who went to the fire.

Mr. —— submitted to the walk with as good a grace as possible. We had a long journey that day, but he manfully accomplished it; and I heard him say, afterwards, that he richly deserved it.

Imagine twenty odd men, 600 miles from the frontiers, at the commencement of a severe winter, deprived of their animals, on an open prairie, surrounded by Camanches, Pawnees and other tribes of hostile Indians. I am fully convinced that but for the "watchfulness" of Col. Fremont, we should have been placed in this awkward predicament.

IMMENSE HERDS OF BUFFALO.

On the divide, between Walnut Creek and the Arkansas River, we travelled through immense herds of buffalo; at one time there could not have been fewer than two hundred thousand in sight.

All around us, as far as the eye could reach, the prairie was completely black with them; they at times impeded our progress. We stopped for more than an hour to allow a single herd to gallop, at full speed, across our path, while the whole party amused themselves with singling out particular ones, and killing them.

I essayed, at different times, to daguerreotype them while in motion, but was not successful, although I made several pictures of distant herds.

On this "divide" I saw numbers of prairie dogs, they ran to their holes on our approach; a small sized owl, most generally stood as sentinel near the hole.

Our Delawares told me that the prairie dog, the owl, and the rattlesnake always congregate together—a strange trio.

The prairie after you pass Pawnee Fork, and also on the divide between Walnut Creek and the Arkansas River, is covered with a short grass, called buffalo grass.

Firewood or timber, only grows on the creek, and the artemisia entirely disappears.

We camped one night on the open prairie, without wood, near Pawnee Fork, a tributary of the Kansas. The thermometer was below freezing point, and there was no vestige of wood or timber to be seen.

I was busily engaged making my daguerreotype views of the country, over which I had to travel the next day. On looking through my camera I observed two of our men approaching over a slope, holding between them a blanket filled with something; curious to know what it was, I hailed them, and found they had been gathering "dried buffalo chips," to build a fire with. This material burns like peat, and makes a very hot fire, without much smoke, and keeps the heat a long time; a peculiar smell exhales from it while burning, not at all unpleasant. But for this material, it would be impossible to travel over certain parts of this immense country. It served us very often, not only for cooking purposes but also to warm our half frozen limbs. I have seen chips of a large size—one I had the curiosity to measure, was two feet in diameter.

Our first camp on the Arkansas was visited by a number of Indian hunters, with the product of their skill, in the use of their bows and arrows, hanging across their horses. One of them borrowed my jack-knife, and cutting a piece of the raw antelope liver, deliberately ate it. I remember the peculiar feeling this exhibition excited in my bosom. I considered the Indian little better than a cannibal, and taking back my knife, turned from him in disgust.

I got bravely over it, however, in the course of my journey, as a perusal of these pages will show.

CHAPTER X.

Cheyenne Indian Village—"Big Timber"—Daguerreotypes—Indian Papoose—Author is Suddenly Changed into a Magician—Silver and Brass Bracelets—Portrait of Indian Princess—"*Presto, pass!*"—Moccasins—Cheyennes and Pawnees at War—Grand Scalp-dance—Col. Fremont a Spectator—Dinner with the Chief—Rosewood Carved Furniture not in Use—Duties of Indian Women—Employment of the Men.

The Cheyenne village, on Big Timber, consists of about two hundred and fifty lodges, containing, probably, one thousand persons, including men, women and children.

I went into the village to take daguerreotype views of their lodges, and succeeded in obtaining likenesses of an Indian princess—a very aged woman, with a papoose, in a cradle or basket, and several of the chiefs. I had great difficulty in getting them to sit still, or even to submit to have themselves daguerreotyped. I made a picture, first, of their lodges, which I showed them. I then made one of the old woman and papoose. When they saw it, they thought I was a "supernatural being;" and, before I left camp, they were satisfied I was more than human.

The squaws are very fond of ornaments; their arms are encircled with bracelets made of thick brass wire—sometimes of silver beaten out as thin as pasteboard. The princess, or daughter of the Great Chief, was a beautiful Indian girl. She attired herself in her most costly robes, ornamented with elk teeth, beads, and colored porcupine quills—expressly to have her likeness taken. I made a beautiful picture of her.

The bracelets of the princess were of brass; silver ones are considered invaluable, and but few possess them.

After I had made the likeness of the princess, I made signs to her to let me have one of her brass bracelets. She very reluctantly gave me one. I wiped it very clean, and touched it with "quicksilver." It instantly became bright and glittering as polished silver. I then presented her with it. Her delight and astonishment knew no bounds. She slipped it over her arm, and danced about in ecstacy. As for me, she thought I was a great "Magician."

My extraordinary powers of converting "brass into silver" soon became known in the village, and in an hour's time I was surrounded with squaws entreating me to make "*presto, pass!*" with their "armlets and brass finger-rings."

Some offered me moccasins, others venison, as payment; but I had to refuse nearly all of them, as I had only a small quantity of quicksilver for my daguerreotype operations.

My "lucifer matches," also, excited their astonishment; they had never seen them before; and my fire water, "alcohol," which I used, also, to heat my mercury—capped the climax.

They wanted me to live with them, and I believe if I had remained, they would have worshipped me as possessing most extraordinary powers of necromancy.

I returned to camp with a series of pictures, and about a dozen pairs of moccasins, some elaborately worked with beads; all of which I stowed away in my boxes, and had the great gratification of supplying my companions with a pair, when they were most required, and when they least expected them.

The Pawnees and Cheyennes were at deadly war, at this time. During our visit to the Cheyenne camp, a number of warriors returned from a successful battle with the Pawnees, and brought in some twelve or fifteen scalps as trophies of their prowess. On the night of their arrival, they had a grand scalp-dance; all the men and most of the women were grotesquely attired in wolf, bear, and buffalo skins; some of them with the horns of the buffalo, and antlers of the deer, for head ornaments. Their faces were painted black and red; each of the chiefs, who had taken a scalp, held it aloft attached to a long pole. An immense fire was burning, around which they danced and walked in procession, while some of the women were beating drums, and making night hideous with their horrible howlings and discordant chantings. This was so novel and extraordinary a scene, that I rode into our camp, about three miles off, and induced Col. Fremont to accompany me to witness it. Mr. Egloffstien, succeeded in writing down the notes of their song; they have no idea of music; they all sing on the same key. I did not notice a single second or bass voice amongst them. We returned to camp about 12 o'clock, and left them still participating in the celebration of their bloody victory. I accepted an invitation to dine with the chief; his lodge is larger, but in no other respects different from those of the others. We dined in it, on buffalo steaks and venison; a fire was burning in the centre; around the fire, were beds made of cedar branches, covered with buffalo robes, on which his two wives and three children slept. They use no furniture of any kind; there are hiding places under their beds, in which they place their extra moccasins and superfluous deer-skin shirts.

The women make the bows and arrows, and all their

moccasins, dress and prepare their skins and buffalo robes, take down and put up their lodges when they move their villages, which is three or four times a year, and all the servile and hard work of the camp. The men hunt, fish, and go to war.

The Cheyennes possess a large number of fine horses, some of which they raise, while the most of them are stolen and taken as prizes in their forays with other tribes of Indians.

CHAPTER XI.

Bent's Trading Post—Purchase Fresh Animals—Buffalo Robes—Immense Lodge—Fremont's Lodge—Doctor Ober—His Scientific Knowledge—Attachment of the Author to him—His Preparation to return to the States—Arkansas River—Giant Cotton Woods—Islands in the Arkansas—Bent's Fort destroyed by Indians—Preparations to cross the Mountains—First View of the Rocky Mountains—Bid adieu to Doctor Ober.

BENT'S HOUSE is built of adobes, or unburnt brick, one story high, in form of a hollow square, with a courtyard in the centre. One side is appropriated as his sleeping apartments, the front as a store-house, while the others are occupied by the different persons in his employ. He has a large number of horses and mules.

Col. Fremont procured from him fresh animals for all the men, leaving behind us those which were thought unable to go through. At this time Mr. Bent had but a small quantity of sugar and coffee; he supplied us, however, with all he could spare, and a considerable quantity of dried buffalo meat, moccasins and overshoes for all the men; a large buffalo-skin lodge, capable of covering twenty-five men, and one small one for Col. Fremont; buffalo robes for each man besides stockings, gloves, tobacco, etc.

I breakfasted with Mr. Bent and Doctor Ober, on baked bread, made from maize ground, dried buf falo meat, venison steaks, and hot coffee; a treat that I had not enjoyed for a very long time.

Col. Fremont having entirely recovered his health, decided not to take the doctor over the mountains, but made arrangements with Mr. Bent to send him home by the first train of wagons; one of our white men, a Mr. Mulligan, of St. Louis, also remained, as an assistant to the doctor. I had formed quite an attachment to Doctor Ober; he was a gentleman of extensive information, and his intellectual capacity was of the highest order. I have ridden by his side for many a mile, listening to his explanations of the sciences of geology and botany. When we passed a remarkable formation, he would stop and compare it with others of similar character in different parts of the world. I regretted very much the necessity there was for his remaining behind, but it was well for him that he did so; his age and make would have incapacitated him from enduring the privations and hardships which we had to encounter.

The weather continuing so cold I found it inconvenient to use my oil colors and brushes; accordingly I left my tin case with the doctor, who promised to take charge of them for me to the States.

When the weather is very clear, you can see the snowy peaks of the Rocky Mountains from Bent's house, which is seventy miles distant. Our friend the doctor wanted to obtain a nearer view of them, and proposed that I should accompany him. We started on a clear morning, for that purpose. I took my apparatus along; we rode thirty miles, but the weather becoming hazy, it entirely shut out our view of the mountains. We returned to camp late at night, after a tiresome day's ride.

EXPEDITION REFITTED.

The Arkansas River where we first struck it, which was at the crossing of the Santa Fé trail, is almost entirely bare of timber; the trunks of several giant cottonwood trees, which had probably been landmarks for early travellers to Mexico, still reared their enormous heads high into the Heavens, defying alike the storms of winter, and the axe and fire of the hungry pioneer, who in vain attempted to hew and burn them down. I measured one of them, its circumference was eighteen feet. We travelled up the river a great many miles, without seeing any timber at all, and relying for firewood on the drift logs, we found along the banks.

There are a great many islands in the Arkansas River, on which some few young cottonwoods are growing. We frequently encamped on these islands.

At "Big Timber," there is a considerable quantity of oak, and cottonwood of large growth. Game of all kinds abounds in it.

Bent's house is a trading post. Indians of the different tribes bring in their venison, buffalo meat, skins, and robes, which are exchanged for various descriptions of manufactured goods. Mr. Bent also receives the annual appropriation from Government, for the neighboring tribes of Indians which are distributed at this point. Bent's Fort, which is situated about thirty miles further up the Arkansas, was recently destroyed by the Indians, and has not been rebuilt, from the scarcity of timber in its vicinity. All the material saved from the fort, was removed to Mr. Bent's house, on Big Timber. After a sojourn of a week, near Bent's trading house, the whole of which time was employed in refitting and preparing proper camp equipage for the journey over the mountains, we bade an affectionate adieu to our worthy doctor; and started in high spirits, the lofty summit of Pike's Peak in the distance glittering in the morning sun.

CHAPTER XII.

Journey up the Arkansas—Bent's Fort—Huerfano River and Valley—Description of the Country—Huerfano Butte—Behind Camp—Daguerreotypes—Scientific Observations—Approach of Night—Trail Lost, and Encampment in the Woods—Buffalo Robes and Blankets—Col. Fremont sends to find us—Bear Hunt—Roubidoux Pass—Emotion of Col. Fremont when Looking upon the Scene of his Terrible Disaster on a Former Expedition—Found a Half Starved Mexican—Col. Fremont's Humanity—His Skill in Pistol Shooting.

WE travelled up the Arkansas, and passing the ruins, of Bent's Fort on the opposite side of the river, struck the mouth of the Huerfano; we followed that river to the Huerfano Valley—which is by far the most romantic and beautiful country I ever beheld. Nature seems to have, with a bountiful hand, lavished on this delightful valley all the ingredients necessary for the habitation of man; but in vain the eye seeks through the magnificent vales, over the sloping hills, and undulating plains, for a single vestige to prove that even the foot of an Indian has ever preceded us. Herds of antelope and deer roam undisturbed through the primeval forests, and sustain themselves on the various cereals which grow luxuriantly in the valley.

But where are the people?

Were there ever any inhabitants in this extraordinarily fertile country?

Will the progress of civilization ever extend so far in the interior?

At present, not even the smoke from an Indian wigwam taints the pure air which plays around, and imparts healthful vigor to my frame.

After crossing the Huerfano River, we saw the immense pile of granite rock, which rises perpendicularly to the height of four or five hundred feet, from a perfectly level valley. It appeared like a mammoth sugar loaf, (called the Huerfano Butte).* Col. Fremont expressed a desire to have several views of it from different distances.

The main party proceeded on the journey, leaving under my charge the mules which carried our apparatus, and also the blankets and buffalo robes of the whole camp; it being necessary in equalizing the weight, to destribute the different boxes on three or four animals. Mr. Egloffstein, Mr. Fuller, and two Delawares, remained with me.

To make a daguerreotype view, generally occupied from one to two hours, the principal part of that time, however, was spent in packing, and reloading the animals. When we came up to the Butte, Mr. Fuller made barometrical observations at its base, and also ascended to the top to make observations, in order to ascertain its exact height. The calculations have not yet been worked out.

If a railroad is ever built through this valley, I suggest that an equestrian statue of Col. J. C. Fremont, be placed on the summit of the Huerfano Butte; his right hand pointing to California, the land he conquered.

When we had completed our work, we found that we were four hours behind camp, equal to twelve miles.

---

* "The Orphan."

We followed the trail of our party, through the immense fields of artemisia, until night overtook us, travelling until we could no longer distinguish the trail.

Our arms were discharged as a signal to the camp; they answered it by firing off their rifles, but the wind being then high, we could not determine their exact distance or position. Then, taking counsel together, we determined to encamp for the night, on the side of a mountain covered with pines, near by.

We soon had a large fire burning, for the weather was intensely cold and disagreeable. Upon unloading our animals we found that we had with us all the baggage and buffalo robes of the camp, but nothing to eat or drink; the night was so dark that although not more than half a mile from a creek, we preferred to suffer from thirst rather than incur fresh danger which might lurk about it.

I had with me three tin boxes, containing preserved eggs and milk, but I preferred to go supperless to bed, rather than touch the small supply which I had, unknown to the rest, carefully hid away in my boxes, to be used on some more pressing occasion.

Our absence was most keenly felt by the camp, for they had to remain up, around their fires all night, not having any thing to sleep on.

We also watched all night, fearful that our animals should stray away, or that we should be attacked by Indians.

At day dawn we reloaded our animals, found our lost trail, and soon met some of our party whom Colonel Fremont had sent to look for us.

When we got to camp, they were all ready for a start, and waiting for us. A delicious breakfast of buffalo and venison had been prepared, and we discussed its merits with an appetite sharpened by a twenty-four hours fast.

At the very base of the Rocky Mountains, while we were approaching the Sand-hill Pass, fresh bear track were discovered by our Delawares, who determined to follow in search of the animal. Diverging a little from our line among the trees on the side of the mountain, our bruin was first seen. "A bear hunt! a bear hunt!" was quickly re-echoed by the whole company. The baggage animals were left to themselves while Colonel Fremont and the whole party darted off at full speed to the chase.

Two of our Delawares who first spied him, were half a mile in advance, for they gave the reins to their animals the instant they saw the bear. His bearship seeing strangers approaching at full speed, and being unused to their ways, thought it most prudent to make himself scarce; he turned and slowly descended the hill in an opposite direction; our loud huzzas finally alarmed him and off he went in full tilt, the whole party surrounding him; the first shot from the Delaware brought him to his knees. Three shots killed him.

He was an enormous black bear, and very fat; 1 partook of but small quantities of it, it being too luscious and greasy for my palate. The meat was brought into camp and served several days for food for the whole party.

The next day I accompanied Col. Fremont into the Roubidoux Pass, from the summit of which I had the first view into the San Louis Valley, the head waters of the "Rio Grande del Norte." On the opposite side forty miles across are the "San Juan Mountains," the scene

COL FREMONT'S BENEVOLENCE.

of Col. Fremont's terrible disaster on a former expedition. He pointed out to me the direction of the spot and with a voice tremulous with emotion, related some of the distressing incidents of that awful night. I made a daguerreotype of the pass with the San Louis Valley and mountains in the distance.

While exploring in the pass we accidentally came upon a Mexican, almost naked, who had deserted or been left behind by some hunters. Col. Fremont, whose great heart beats in sympathy for the suffering of his fellow men, made him follow to camp, and although he knew that this man would be an incubus upon the party from his inability to walk, allowed him to accompany the expedition, and supplied him with a part of his own wardrobe. This man subsequently proved perfectly worthless.

On our way down from the pass, Col. Fremont took out his revolver, and at a distance of about twenty paces killed a small, white, delicately formed animal, very like an ermine. This was an excellent shot with a sightless pistol.

CHAPTER XIII.

Sand-hill Pass—San Louis Valley—Natural Deer-park—Smoked Venison—Last sight of Game—Rio Grande del Norte—Sarawatch—Cochotope Pass—First Snow in Mountains—Gunnison's Wagon Trail—Summit of Pass—Waters commence to flow towards the Pacific—Encampment—Immense Rugged Mountain—Impracticability of ascent by Mules—The Author ascends on Foot—Col. Fremont accompanies him—Daguerreotype Panorama from its Summit—Col. Fremont's Consideration for his Men—Sublimity—First View of Grand River—Reflections—Return to Camp.

WE entered the San Louis Valley through the Sand-hill Pass, and camped at the mouth. Travelling up the valley about twenty miles, we ascended one of the verdant and gentle slopes of the mountains, along which meandered a stream of living water, fringed on its banks with cottonwood and elms. We selected a camp-ground in an immense natural deer-park, and raised our tents under the shelter of wide-spreading cedars.

Scarcely were we comfortably fixed, when a herd of black-tail deer came down the mountain to water within sight of our camp. Cautiously our Indian hunters sallied out, and ere many minutes, the sound of one, two, three—a dozen rifles were heard in quick succession. Every shot brought down a fine fat buck, and our supper that night, consisted of as fine roast venison as ever graced the table of an epicure.

Col. Fremont determined to remain here for several days in order to have a quantity of the meat cured for our use in the mountains. I exercised my skill in rifle shooting for the last time at this camp. Game of all kinds which had hitherto been plentiful, disappeared almost entirely after we left it.

We travelled up the San Louis Valley, crossing the Rio Grande del Norte, and entered the Sarawatch Valley through a perfectly level pass. Our journey continued along the valley until we came to the Cochotope, where we camped.

That night it snowed on us for the first time. The snow obliterated the wagon tracks of Capt. Gunnison's expedition, but Col. Fremont's unerring judgment conducted us in the precise direction by a general ascent through trackless, though sparsely timbered forests, until we approached the summit, on which grew an immense numbers of trees, still in leaf, with only about four inches of snow on the ground.

As we approached this dense forest, we soon perceived that the axe of the white man had forced a passage through for a wagon-road. Many of the larger trees on both sides of the track were deeply cut with a cross, as an emblem of civilization, which satisfied us that Capt. Gunnison and Lieut. Beale had penetrated through to the other side. In this forest, we were surrounded by immense granite mountains, whose summits were covered probably with everlasting snow. The streams from them which had previously been running towards us, now took the opposite direction, supplying us with the gratifying proof that we had completed our travel to the summit, and were now descending the mountains towards the Pacific. After issuing from these woods we camped on the edge of a rivulet.

At this camp Col. Fremont exhibited such unmistakable marks of consideration for me, that it induced

4*

my unwavering perseverance in the exercise of my professional duties subsequently, when any other man would have hesitated, and probably given up, and shrunk dismayed from the encounter.

Near by our camp, a rugged mountain, barren of trees, and thickly covered with snow, reared its lofty head high in the blue vault above us. The approach to it was inaccessible by even our surefooted mules. From its summit, the surrounding country could be seen for hundreds of miles. Col. Fremont regretted that such important views as might be made from that point, should be lost, and gave up the idea as impracticable from its dangerous character. I told him that if he would allow two men to assist me in carrying my apparatus up the mountain, I would attempt the ascent on foot, and make the pictures; he pointed out the difficulties, I insisted. He then told me if I was determined to go he would accompany me; this was an unusual thing for him and it proved to me, that he considered the ascent difficult and dangerous, and that his superior judgment might be required to pick the way, for a misstep would have precipitated us on to the rugged rocks at its base; and it also proved that he would not allow his men or officers to encounter perils or dangers in which he did not participate.

After three hours' hard toil we reached the summit and beheld a panorama of unspeakable sublimity spread out before us; continuous chains of mountains reared their snowy peaks far away in the distance, while the Grand River plunging along in awful sublimity through its rocky bed, was seen for the first time. Above us the cerulean heaven, without a single cloud to mar its beauty, was sublime in its calmness.

Standing as it were in this vestibule of God's holy Temple, I forgot I was of this mundane sphere; the divine part of man elevated itself, undisturbed by the influences of the world. I looked from nature, up to nature's God, more chastened and purified than I ever felt before.

Plunged up to my middle in snow, I made a panorama of the continuous ranges of mountains around us. Col. Fremont made barometrical and thermometrical observations, and occupied a part of his time in geological examinations. We descended safely, and with a keen appetite, discussed the merits of our dried buffalo and deer meat.

CHAPTER XIV.

Intense Cold—Author's First Journey on Foot—Immense Mountains of Snow—Escape of his Pony—Lose Sight of Companions—Arrival at top of the Mountain—Pony Recovered—Revolution of Feeling—Indian Gratitude Exemplified—Horse Steaks Fried in Tallow Candles—Blanc Mange—New Year's Day—Dangerous Ascent of a Mountain—Mules tumble Down—Animals Killed—Successful Attempt Next Day —Camp in four feet of Snow—Coldest Night—Sleep out in open Snow.

EATING, sleeping, and travelling, continually in the open air, with the thermometer descending, as we gradually ascended the immense slopes of country between the frontiers of Missouri and the Rocky Mountains, until I have found myself in a temperature of 30° below zero, prepared my system for the intense cold, which we endured during our journey through that elevated country. Twice only did our party find it too cold to travel longer than half an hour, without stopping and making a large fire to keep ourselves from freezing. We were all mounted at the time, but we found it necessary to walk a greater part of the way, to keep up a circulation of the blood.

It is judiciously ordained by a kind Providence, that the cold as well as heat, gradually increases in intensity.

If the human body at a temperate heat, say 80° was suddenly exposed to a temperature of 30° below zero, in which we travelled without any extra clothing, no ill effect resulting, we should not have been able to exist for an hour.

Let us then humbly acknowledge that to the great Omnipotent, we owe our being and all the benefits we receive.

MY FIRST JOURNEY ON FOOT.

It was a very cold day in December; the snow covered the immense mountain, over which we had to travel, and right merrily we all followed each other's footsteps in the deep snow.

When we arrived at the foot of the rugged mountain, it was found necessary to dismount, and lead our animals along the intricate and tortuous path. As usual I was at the rear of the cavalcade; I threw the bridle over my pony's head, and followed slowly behind him. I plunged frequently up to my neck in chasms of snow. My efforts to extricate myself cost me some time, and when I regained my footing, I discovered my pony about fifty yards ahead, trying to regain the party. I redoubled my exertions to reach him—I hallooed all to no purpose—I sank down exhausted on a rock, with the dreadful reality that I was alone, and on foot on the mountains of eternal snow, with a long day's journey before me.

Gathering fresh strength and courage from the serious position I found myself in, I scrambled up that mountain with a heart palpitating so loudly, that I could count its pulsations. In this manner, alternately resting, I reached the top. On looking on the other side, the only indication of the party, was their deep trail in the frozen snow.

I commenced descending, and at considerable distance below me, I fancied I saw a moving object under a tree; continuing in the track, slipping at times a distance of ten or fifteen feet, until some disguised rock brought me up, I reached the bottom, where I found my pony tied to a tree, immediately on the trail.

No shipwrecked mariner on beholding the approach of a friendly vessel to deliver him from certain death, ever felt greater joy than I did, when I realized that it was my horse which I saw.

This incident was most injurious to me, and I felt its effects for several days, both in body and mind. I mounted my pony, and arrived in camp at dark, some four or five hours after the rest of the party.

Captain Wolff saw my pony riderless, and suspecting that he had escaped from me, caught and tied him up in the place where he was sure to be found; thus repaying me a hundred fold for my medical advice and attendance on Salt creek.

* * * * * *

HORSE STEAKS FRIED IN TALLOW CANDLES, AND BLANC MANGE FOR DESSERT.

At Bent's Fort, Col. Fremont had several pounds of candles made out of buffalo tallow; the want of convenient boxes to convey them, resulted in many of them being broken to pieces, so as to render them useless as candles. On the first of January, 1854, our men were regaled by unexpected, though not unwelcome luxuries.

I had reserved with religious care, two boxes containing one pound each, of Alden's preserved eggs and milk.—(The yolks of the eggs were beaten to a thick paste with a pound of loaf sugar, the milk was also prepared with powdered sugar, and hermetically sealed in tin cases.)—These two tins I had stowed away in my boxes, being the remains of the six dozen which had been wantonly destroyed at our six weeks camp on Salt Creek.

Nobody knew I had them. A paper of arrow root, which my wife had placed in my trunk, for diet, in case I was sick, I had also reserved. These three comestibles, boiled in six gallons of water, made as fine a blanc mange as ever was *mangéd* on Mount Blanc. This "dessert" I prepared without the knowledge of Col. Fremont.

Our dinner, in honor of "New Year's Day," consisted, besides our usual "horse soup," of a delicious dish of horse steaks, fried in the remnants of our "tallow candles." But the satisfaction and astonishment of the whole party cannot be portrayed, when I introduced, as dessert, my incomparable blanc mange. "Six gallons of *bona fide*," nourishing food, sweetened and flavored! It is hardly necessary to say, that it disappeared in double quick time. The whole camp had a share of it; and we were all sorry that there was "no more left of the same sort."

*   *   *   *   *

Several days after we came down from the Cochotope Pass, it became necessary to ascend a very high and excessively steep mountain of snow. When we were half way up, one of the foremost baggage mules lost his balance, from his hind feet sinking deep in the snow. Down he tumbled, heels-over-head, carrying with him nearly the whole cavalcade, fifty odd in number, several hundred feet to the bottom.

It was a serious, yet a most ludicrous spectacle, to witness fifty animals rolling headlong down a snow mountain, gaining fresh impetus as they descended, unable to stop themselves. The bales of buffalo robes, half buried in the snow, lodged against an old pine tree, the blankets scattered everywhere; my boxes of daguerreotype materials uninjured, although buried in the snow. Considerable time was occupied in searching after them.

I found myself standing up to my eyes in snow, high up the mountain, witnessing this curiously interesting, although disastrous accident; for, when we collected ourselves and animals together, we found that one mule and one horse were killed. This scene made a deep impression upon me. Night came upon us before we were ready to leave the spot. We camped on the same place of the night before.

A snow storm commenced raging, which detained us in this situation for another day; when, determined to cross the mountain, we all recommenced the ascent, and successfully arrived, though much exhausted, without further accident, at the top, and encamped on its summit in snow four feet deep.

That night the thermometer sank very low, and the men stood to their waists in snow, guarding the animals to prevent their running away in search of grass, or something to eat.

We descended the mountain the next day. Our tent poles, belonging to the large lodge, were broken by their contact with the trees in the winding path. The lodge, afterwards, became useless, and the men, myself among them, had to sleep out upon the open snow, with no covering but our blankets, etc.

CHAPTER XV.

Descent of Snow Mountains—Gun for a Walking-stick—Indian Tracks—Examination of Arms—Predicament of the Author—Lecture from Col. Fremont—Wild Horse Killed by Indians—Utah Indian Village—Encampment—Trade for Venison—Camp at Night Surrounded by armed Indians—They Demand Payment for the Horse Killed by the Indians—Col. Fremont's Justice—Indians want Gunpowder—Their Demand Refused—Massacre of the Party Threatened—Defiance—Pacification—Author Trades for a Horse—He Leaves his Colt's Revolver in Camp—Runaway Horse—Author Finds himself in a Sage Bush—Pistol Recovered—Trouble in Perspective—Exchanges Horses—Lame Horse—Author on Foot—Regrets that he was not Educated for a Horse-Breaker.

After descending a very steep mountain, on the snows of which we passed the coldest night I experienced during the journey, the thermometer, at daylight, being 30 degrees below zero, we camped on a creek fringed with willows and interspersed with cotton-wood. The country indicating that there might be game about, our Delawares sallied out in quest of some.

We at this time were on rations of meat-biscuit,* and had killed our first horse for food. Towards night, our hunters returned, and brought with them the choice parts of a fine fat, young horse that they had killed. He was one of three or four wild ones which they discovered grazing some four miles from camp.

Our men, in consequence, received a considerable addition to their stock of provisions, which, when cooked,

* A preparation made by saturating flour with the juices of boiled beef, and then baked into biscuit.

proved much more palatable than our broken down horses.

The Delawares also discovered recent footprints of Utah Indians. This information caused Col. Fremont to double the guard and examine the arms of the whole party, who hitherto had been warned by him of the necessity there was for keeping them in perfect order.

Suddenly it occurred to me that my double-barrel gun might be out of order: I had used it as a *walking-stick*, in descending the mountain that day; the snow was so deep that I was obliged to resort to that course to extricate myself from the drifts.

I quietly went to the place where I had laid it down, and attempted to fire it off; both caps exploded, but the gun did not go off, the barrels being filled with frozen snow. The quick ear of Col. Fremont heard the caps explode. He approached me very solemnly and gave me a lecture, setting forth the consequences which might have ensued from a sudden attack of the Indians on our camp. "Under present circumstances, Mr. Carvalho," said he, "I should have to fight for you." His rebuke was merited, and had its effect throughout the camp, for all the men were most particular afterwards in keeping their arms in perfect order.

We travelled that day nearly twenty miles, and encamped outside of a Utah Indian village, containing a large number of lodges and probably several hundred persons.

The men were mostly armed with rifles, powder-horns, and also with their Indian implements of warfare. On our mules was packed the balance of our "fat horse" of the night before.

These Indians received us very kindly, and during the

evening we exposed our wares, viz.: blankets, knives, red cloth, vermilion, etc., etc., which we brought along to conciliate the Indians, and also to trade with them for horses and venison.

We made several purchases, and traded for several small lots of fat venison.

About nine o'clock, after placing double guard around our animals and while we were regaling on fat deer meat in Col. Fremont's lodge, we heard loud noises approaching the camp; voices of women were heard in bitter bewailment. I thought it was a religious ceremony of Indian burial, or something of the kind. Col. Fremont requested me to see from what it proceeded. I found the whole Indian camp in procession assembled around our lodge. The warriors were all armed, headed by a half-breed, who had been some time in Mexico, and had acquired a smattering of the Spanish language; this man acted as interpreter. Understanding the Spanish language, I gleaned from him that the horse our Delawares had killed the evening before, some twenty miles away, belonged to one of the squaws then present, who valued it very highly, and demanded payment.

On informing Col. Fremont, who denied himself to the Indians, he remarked that "we had no right to kill their horse without remunerating them for it." The man in charge of the baggage was deputed to give them what was a fair compensation for it.

The Indians having seen our assortment, wanted a part of everything we had, including a keg of gunpowder.

To this demand Col. Fremont gave an absolute refusal, and at the same time emphatically expressed his desire that the men should not sell, barter, or give away a single grain of gunpowder, on pain of his severest displeasure.

The Indians then threatened to attack us. Col. Fremont defied them. After considerable parleying, we succeeded in pacifying them.

As it was the intention of Col. Fremont to leave camp at an early hour, I unpacked my daguerreotype apparatus, at daylight, and made several views.

While engaged in this way, one of the Utah Indians brought into camp a beautiful three-year-old colt, and offered to trade him with me; he was a model pony—dark bay color, in splendid order, sound in wind and limb, and full of life and fire. My poor buffalo Pungo had, three days before, been shot down for food, and in consequence I was literally on foot, although I was using one of the baggage animals for the time.

With permission of Col. Fremont I traded for him; I gave him in exchange one pair of blankets, an old dress coat, a spoiled daguerreotype plate, a knife, half an ounce of vermilion and an old exhausted pony, which we would have been obliged to leave behind; previous to the trade, I had never mounted him, but I saw the Indian ride him, and his movements were easy and graceful. The Indian saddled him for me, as I was otherwise engaged, and did not notice him during the operation. By this time the rest of the party were all mounted, and I never jumped on him until the last moment; he winced a little under the bit, the first one he ever had in his mouth, but cantered off at a round pace, I would not at that moment have taken $500 for him. I considered myself safely mounted for the rest of the journey.

After we proceded about two miles, my pony prancing and caracoling to the admiration of the whole party, I discovered that I had left my Colt's navy revolver in camp. I told Col. Fremont of my carelessness, and he smilingly sent one of the men back with me to look for it. I must confess I had not the slightest hopes of finding it, nor had he.

At the time we started, there must have been two hundred Utah men, women, and children at our camp, and if one of them had picked it up, it was most unlikely I should ever receive it again. They had shown some hostility, and although I was not afraid to go back, I thought some danger attended it—Frank Dixon accompanied me.

My pony finding his head turned homeward, commenced champing at his bit, and working his head and body endeavoring to get away. I prided myself on being a good horseman, but this fellow was too much for me.

He got the bit between his teeth and off he started at a killing pace for camp. In less than five minutes I found myself in a wild sage bush on the road; the saddle had slipped round his body, which was as smooth as a cylinder, while I, losing my balance, slipped off.

My pony was quietly grazing in the Indian camp, when I, riding double with Frank, arrived there. The most important thing, was my pistol; I proceeded immediately to the spot, and, hidden in the long grass, where I laid it down, I found it.

With the assistance of the Utahs, my pony was captured, and doubling the saddle-blanket, I attempted to draw the girth tightly—he resisted, and gave considerable trouble; but I was finally mounted, and away we cantered after our party, which we overtook after a couple of hours' ride.

This animal continued to trouble me every morning afterwards. On one occasion, I was saddling him, to perform which operation, I had to tie him to a tree, if one was at hand; at the time I now describe, he was tied to a tree, and in vain I endeavored to place the saddle on him, finally, he reared, and planted both feet on my breast, and I barely escaped with my life, yet my pride never suffered me to complain about it. Sometimes one of my comrades would assist me, but on this occasion, Col. Fremont saw my predicament; in a few minutes, his servant, "Lee," came to me, and said, "he was more accustomed to break horses than I was," and offered to exchange with me, until mine was more manageable.

This man rode a cream colored pacer, which Col. Fremont wanted to take through to California, if possible, as a riding horse for his daughter. I need not say how gladly I accepted this offer. I rode out of camp that morning much lighter in spirits, although suffering somewhat from the bruises I received. The horse I exchanged for, was a pacer, he had no other gait; and unaccustomed to it, I did not notice, until one of the Delawares pointed out to me, that there was any defect in him.

Captain Wolff was riding by my side during the day, and expressed in his Indian manner, how surprised he was that I had exchanged my fresh pony for a lame pacer, "one day more, that horse no travel, Carvalho go foot again!"

His prognostications proved, alas, too true, for on the second day, he was so lame that I could not ride him, and I remained on foot, while my beautiful pony was gallantly bearing the cook.

The horse, he said, was not lame when he gave him to me, and I could not prove that he was, so I was constrained to submit, but I never saw this man galloping past me, while I was on foot, that I did not regret I was not brought up as an " ostler and professional horsebreaker."

CHAPTER XVI.

Grand River—Descent of Mounted Indians into Camp—Military Reception—Their demands—Trouble Expected—Excitement of the Author—Exhibition of Colt's Revolvers—Col. Fremont's Knowledge of Indian Character—The Great Captain in his Lodge—Alarm of the Indians—Quadruple Guard—Departure of Indians—Vigilance the price of Safety—Crossing of the Grand River—Horse Killed for Food—Review of Our Position—Impressive Scene—Cold Night—Mr. Fuller—Whites without Food—Beaver Shot—The Camp under Arms—False Alarm.

When we left the Utah village, we travelled a long day's journey, and camped on the Grand River, thirty miles from the last camp; my pony behaved admirably well on the road, and I would not have parted with him on any account.

While at supper, the guard on the look-out gave the alarm that mounted Indians were approaching, the word was given to arm and prepare to receive them.

About fifty or sixty mounted Utah Indians, all armed with rifles, and bows and arrows, displaying their powder horns and cartouch boxes most conspicuously, their horses full of mettle, and gaily caparisoned, came galloping and tearing into camp.

They had also come to be compensated for the horse we had paid for the night before; they insisted that the horse did not belong to the woman, but to one of the men then present, and threatened, if we did not pay them a great deal of red cloth, blankets, vermilion, knives, and gunpowder, they would fall upon us and massacre the whole party.

On these occasions, Col. Fremont never showed himself, which caused the Indians to have considerable more respect for the "Great Captain," as they usually called him; nor did he ever communicate directly with them, which gave him time to deliberate, and lent a mysterious importance to his messages. Very much alarmed, I entered Col. Fremont's lodge, and told him their errand and their threats. He at once expressed his determination not to submit to such imposition, and at the same time, laughed at their threats; I could not comprehend his calmness. I deemed our position most alarming, surrounded as we were by armed savages, and I evidently betrayed my alarm in my countenance. Col. Fremont without apparently noticing my nervous state, remarked that he knew the Indian character perfectly, and he did not hesitate to state, that there was not sufficient powder to load a single rifle in the possession of the whole tribe of Utahs. "If," continued he, "they had any ammunition, they would have surrounded and massacred us, and stolen what they now demand, and are parleying for."

I at once saw that it was a most sensible deduction, and gathered fresh courage. The general aspect of the enemy was at once changed, and I listened to his directions with a different frame of mind than when I first entered.

He tore a leaf from his journal, and handing it to me, said: here take this, and place it against a tree, and at a distance near enough to hit it every time, discharge your Colt's Navy six shooters, fire at intervals of from ten to fifteen seconds—and call the attention of the Indians to the fact, that it is not necessary for white men to load their arms.

I did so; after the first shot, they pointed to their own rifles, as much as to say they could do the same, (if they had happened to have the powder), I, without lowering my arm, fired a second shot, this startled them.

I discharged it a third time—their curiosity and amazement were increased: the fourth time, I placed the pistol in the hands of the chief and told him to discharge it, which he did, hitting the paper and making another impression of the bullet.

The fifth and sixth times two other Indians discharged it, and the whole six barrels being now fired it was time to replace it in my belt.

I had another one already loaded, which I dexterously substituted, and scared them into an acknowledgment that they were all at our mercy, and we could kill them as fast as we liked, if we were so disposed.

After this exhibition, they forgot their first demand, and proposed to exchange some of their horses for blankets, etc.

We effected a trade for three or four apparently sound, strong animals; "Moses," one of the Delaware chiefs, also traded for one, but in a few days they all proved lame and utterly useless as roadsters, and we had to kill them for food.

The Indians with the consent of Col. Fremont, remained in camp all night; they had ridden thirty miles that day, and were tired. On this occasion, eleven men, fully armed, were on guard at one time.

The Indians who no doubt waited in camp to run our horses off during the night, were much disappointed in not having an opportunity. They quietly departed the next morning, while our whole camp listened to the energetic exclamation of Col. Fremont, that the "price of safety is eternal vigilance."

The crossing of the Grand River, the eastern fork of the Colorado, was attended with much difficulty and more danger. The weather was excessively cold, the ice on the margin of either side of the river was over eighteen inches thick; the force of the stream always kept the passage in the centre open; the distance between the ice, was at our crossing, about two hundred yards. I supposed the current in the river to run at the rate of six miles an hour. The animals could scarcely keep their footing on the ice, although the men had been engaged for half an hour in strewing it with sand. The river was about six feet deep, making it necessary to to swim our animals across; the greatest difficulty was in persuading them to make the abrupt leap from the ice to the roaring gulph, and there was much danger from drowning in attempting to get on the sharp ice on the other side, the water being beyond the depth of the animals, nothing but their heads were above water, consequently the greater portion of their riders' bodies were also immersed in the freezing current.

To arrive at a given point, affording the most facilities for getting upon the ice, it was necessary to swim your horse in a different direction to allow for the powerful current. I think I must have been in the water, at least a quarter of an hour. The awful plunge from the ice into the water, I never shall have the ambition to try again; the weight of my body on the horse, naturally made him go under head and all; I held on as fast as a cabin boy to a main-stay in a gale of wind. If I had lost my balance it is most probable I should have been drowned. I was nearly drowned as it was, and my clothes froze stiff upon me when I came out of it. Some of the Delawares crossed first and built a large fire on the other side, at which we all dried our clothes standing in them.

It is most singular, that with all the exposure that I was subjected to on this journey, I never took the slightest cold, either in my head or on my chest; I do not recollect ever sneezing. While at home, I ever was most susceptible to cold.

The whole party crossed without any accident; Col. Fremont was the first of our party to leap his horse into the angry flood, inspiring his men, by his fearless example to follow.

"Julius Cæsar crossed the Rubicon with an immense army; streams of blood followed in his path through the countries he subdued, to his arrival at the Eternal City, where he was declared dictator and consul."

On a former expedition, Col. Fremont crossed the Grand River with a handful of men; but no desolation followed in his path. With the flag of his country in one hand and the genius of Liberty resting on his brow, he penetrated through an enemy's country, converting all hearts as he journeyed, conquering a country of greater extent than Cæsar's whole empire, until he arrived at San Francisco, where he became military commandant and governor in chief of California, by the simple will of the people. Fremont's name and deeds, will become as imperishable as Cæsar's.

At last we are drawn to the necessity of killing our brave horses for food. The sacrifice of my own pony that had carried me so bravely in my first buffalo hunt, was made; he had been running loose for a week unable to bear even a bundle of blankets. It was a solemn event with me, and rendered more so by the impressive scene which followed.

Col. Fremont came out to us, and after referring to the dreadful necessities to which we were reduced, said "a detachment of men whom he had sent for succor on a former expedition, had been guilty of eating one of their own number." He expressed his abhorrence of the act, and proposed that we should not under any circumstances whatever, kill our companions to prey upon them. "If we are to die, let us die together like men." He then threatened to shoot the first man that made or hinted at such a proposition.

It was a solemn and impressive sight to see a body of white men, Indians, and Mexicans, on a snowy mountain, at night, some with bare head and clasped hands entering into this solemn compact. I never until that moment realized the awful situation in which I, one of the actors in this scene, was placed.

I remembered the words of the sacred Psalmist, (Psalm cviii. 4–7) and felt perfectly assured of my final deliverance.—"They *wandered* in the *wilderness* in a solitary way: They found no city to dwell in.

"*Hungry* and *thirsty* their souls fainted within them. Then they cried unto the Lord in their trouble, and he *delivered them* out of their distresses.

"And he led them forth by the *right way* that they might go to a *city of habitation.*

"Oh, that *men* would *praise the Lord* for his goodness, and for his wonderful works to the children of men."

It was a clear, cold night, on the Eagle Tail River, after a long fast, and a dreary walk, our men had returned supperless to sleep on their snowy bed, and with no prospect of anything to eat in the morning, to refresh them for another day's tramp. It was a standing rule in camp that a rifle discharged between the set of watch at night until daylight, was a signal that Indians were approaching, and this rule had been strictly observed, as a safeguard to the party. I have seen our camp on Salt Creek surrounded with wolves—they even came within its precincts and stole our buffalo meat, but our Delawares would never allow an arm to be discharged. On this occasion, Mr. Fuller was on guard, and it was a few days before he gave out. We had been twenty-four hours without a meal, and as may be supposed, he was as hungry as the rest of us; while patrolling up and down the river on the banks of which we were encamped, his keen eye discovered a beaver swimming across the stream; he watched it with rifle to his shoulder, and as it landed, he fired and killed it.

The sudden discharge of a rifle during a still night, under overhanging mountains, and in the valley of the river where we expected to find Indians, made a tremendous explosion. The sound reverberated along the rocks, and was re-echoed by the valley. Instantly the whole camp was on duty. Col. Fremont who had been making astronomical observations, had but a few moments previously retired to rest. He rushed out of his lodge, completely armed, the party assembled around it and all were filled with the utmost anxiety and alarm. We did not know the number or character of the enemy, but we were all prepared to do battle to the death. In a few moments, one of the Delawares approached camp dragging after him an immense beaver, which he said Mr. Fuller had killed for breakfast. The sight of something to eat, instead of something to fight,

A FALSE ALARM. 103

created quite a revolution of feeling; and taking into consideration the extremity, which caused Mr. Fuller to break through the rule, Col. Fremont passed it off quietly enough. Poor Fuller did not realize the excited condition of the camp, until he was relieved from duty. Our beaver was dressed for breakfast, when Fuller told Col. Fremont that he was so anxious and delighted at seeing the beaver, that he entirely forgot the rule of the camp.

CHAPTER XVII.

Divide between Grand and Green River—Capt. Gunnison's Trail—Without Water—Formation of the Country—Castellated Bluffs—Green River Indians—Crossing of the Green River—Interview with Indians—Disappointment—Grass-seed—Manner of Preparing it for Food—Horse Purchased—Starving Condition of the Whites—Incident Exhibiting the Moral Dishonesty of one of the Men—Name not Published—Dinner on Porcupine—"Living Graves"—Tempestuous Night—Reflections on Guard—No Grass—Frozen Horse Liver—Blunted Feelings.

THE divide between Grand River and the Green River, (the eastern and western forks of the Colorado) is barren and sterile to a degree. At the season that we crossed, there was no water between the two rivers, a distance of about forty miles. Capt. Gunnison's wagon trail was still plainly visible at the crossing of a gully, now however without water.

That party must have had great difficulty in transporting their wagons across it. From its appearance, a tremendous body of water must have forced a passage through the gully, at that time. Dwarf artemisia grows sparsely on this sandstone formation.

At the roots of the artemisia still remained small quantities of dry powdered snow. To allay my thirst, I have put my head under the bush, and lapped the snow with my tongue. The descent into the valley of the Green River was over most dangerous projections of different strata of rock, thrown into its present state by some convulsion of nature.

GRASS-SEED FOR FOOD.

When we arrived at the river, we saw on the high sand bluffs, on the opposite side, several Indians, whose numbers soon increased. As our party was much exhausted for want of wholesome food, we were buoyed up with hopes that we could obtain supplies from them.

We crossed the river, and were conducted by the Indians to a fertile spot on the western bank of it, where their village was. We found that they lived on nothing else but grass-seed, which they collected in the fall. Their women parch it, and grind it between stones. In this manner it is very palatable, and tastes very much like roasted peanuts.

This, their only article of food, was very scarce, and we could procure only a small supply. I parted with everything out of my daguerreotype boxes that I did not require, and several articles of necessary clothing, for about a quart of it. It is very nourishing, and very easy of digestion. The quantity I had, lasted me for three days. I made a hearty meal of it the night we camped among them.

To the sustaining proporties of this cereal, I firmly believe, I owe the strength which enabled me to undergo the physical exertion that was required to reach the settlements.

Each man procured a more or less quantity.

Col. Fremont purchased a lame horse, in very good condition, which was slaughtered at this camp; and an incident occurred which proved to me the real character of one of my companions.

At the killing of this horse, nearly all the men were present. They had not tasted food for nearly two days, and were, consequently, ravenous, and thought of nothing else but satisfying the cravings of hunger

As soon as the horse was slaughtered, without exception, every one cut off a piece, and roasted it at the different camp fires. This was contrary to camp discipline; and, a complaint was made to Col. Fremont, by one of the Delawares, of what was going on, Mr. —— was among the first to cut off pieces from the meat, and he devoured larger quantities than the rest of us. When Col. Fremont was approaching, he took his pencil and paper out of his pocket, and seating himself by the fire, appeared to be deeply absorbed in his occupation. The rest of us remained where we were, partaking of the roast. Col. Fremont lectured us all for not waiting until supper, to eat our respective shares, and pointed this "gentleman" out as an exception, and as one who exercised "great self-denial." At the same moment, he had a piece of meat, *covered up in the cinders, at his feet!*

This "gentleman," instead of avowing his complicity, encouraged the mistake of Col. Fremont, by his continued silence. If he ever reads this journal, he will recognize himself, and, probably, not thank me for withholding his "name" from the public.

One of the most tiresome and unpleasant of duties devolves on those of the party who are at the end of the cavalcade. This duty is, driving up the animals which, either from exhaustion or other causes, linger on the road. Stopping on the trail to make daguerreotypes, generally placed me in the rear; and I have often overtaken the muleteers with a dozen lazy or tired animals, using, in vain, all their endeavors to make them go ahead. As a rule, I always assisted them, sometimes on foot, and in the earlier part of the journey on horseback. When a mule takes a stand, and deter mines not to budge a step, it requires a man with an extraordinary stock of patience to wait upon his muleship's leisure.

The idea frequently suggested itself, that I should change my professional card-plate, and add instead, *my name with* "*M. D.*" attached, as significant of my new office.

DINNER ON PORCUPINE.

A large porcupine was killed and brought into camp to-day by our Delawares, who placed it on a large fire burning off its quills, leaving a thick hard skin, very like that of a hog. The meat was white, but very fat, it looked very much like pork. My stomach revolted at it, and I sat hungry around our mess, looking at my comrades enjoying it. The animal weighed about thirty pounds.

RELIEVE THE GUARD.

I was awakened one night by a rude push from the officer of the guard, who was a huge "Delaware." "Carvalho, go watch horse." "Twelve o'clock." I put my head out of my buffalo robe, and received a pile of fresh snow upon me. I had laid myself down on a snowbank, before a scanty fire of artemisia. I had my clothes on, and wrapped in my buffalo robe, I had sought a few hours sleep until my turn to guard arrived.

I came into camp exhausted, from a ten mile travel on foot, over an irregular and broken road. I had stopped to make daguerreotypes; in consequence, I was detained, and did not get to camp until near eight o'clock.

With some difficulty I threw off the heavy snow which enveloped me, and soon discovered that a northeast snow storm was furiously raging. The fire was extinguished, and six inches of snow now lay on the ashes. I took hold of my gun from under my buffalo robe, and asked the Delaware, "where the animals were."

He pointed in the direction, and replied,—" horses on the mountain, one mile away." I looked out, but could not see ten feet ahead. I thought of the remark my good old mother made on a less inclement night, when I was a boy, and wanted to go the play. " I would not let allow a cat to go out in such weather, much less my son."

Dear soul! how her heart would have ached for me, if she had known a hundredth part of my sufferings.

I followed in the direction given me, and succeeded in finding the animals. I relieved my companion, and walking in snow up to my waist, around the animals for two hours, formed my sole occupation.

There was no grass. The horses and mules were hungry, and whenever they could steal a chance, they would wander out of the corral, and give us trouble to hunt them back; on this night they were very restless, and gave the guard continual exercise, which was also necessary to keep the life within them; it was comparatively easy to walk around in the track; but when one went astray, every step you took, plunged you two feet deep in the snow, making it a most tiresome and arduous task. The two hours seemed at least six, before I was relieved, when groping my way down the mountain side, I followed the trail to camp; by this time the last guard had made a fresh fire of artemisia, which consumes quickly, and burns brightly while consuming. I laid on a fresh pile, and by its light I saw the living graves of my companions; there they lay, with snow underneath them for a bed, and the "cold mantle of death," as it were, above them for a coverlid.

Cold, tired and hungry, I rested myself before the fire, and warmed my frozen limbs.

Some little distance from the fire, now covered with snow, lay the frozen meat of the horse we had killed the night before; all in the camp were fast wrapped in sleep. I was the only one awake. Taking out my jack-knife, I approached the pile of meat intended for the men's breakfast, and cutting about a half pound of the liver from it, I returned to the fire, and without waiting to cook it, I consumed it raw—the finer feelings of my nature were superseded by the grosser animal propensities, induced most probably from the character of the food we had been living upon for the last forty days.

I filled my pipe, and sat wrapped in my robe, enjoying the warmth of my fire, determined to remain by it until my tobacco was consumed.

The wind, which had been blowing from the N.E., now chopped round to the N.W., dissipating the snow-clouds. The glorious queen of night shone forth in resplendent brilliancy. With the change of wind came an increase of cold—the thermometer, at daylight that morning indicated 20° below zero. One of my feet which was much blistered became numbed, and gave me intense pain. I took off my moccasin, and rubbing my foot in the snow to create circulation, I partially relieved it.

Finding it more comfortable, lying down, I crept

under the snowy robe, and made the comparison of the warm rooms, feather beds, and silken canopies of the St. Nicholas wedding-chamber, with our snow-wreathed pillows, airy rooms, and the starry canopy of heaven.

CHAPTER XVIII.

Careless Packing of Animals—Mule Missing—Their value as Roadsters—Col. Fremont's Horse gives out—His Humanity Exemplified—Wolf killed for Food—Raven Shot—River Bottom—Original Forest—Large Camp Fires—Terrible Rain Storm—Disagreeable Bed—Darkness—Fires Extinguished—Value of Rain—Glorious Sunrise—Contrast with Home Comforts.

CARELESS PACKING OF THE MULES.

FROM careless packing of the mules, many of our party were often detained on the road. A bale of blankets or buffalo robes, would be displaced while descending some steep mountain; the mule, finding himself free from his load, would dart off in an opposite direction at full speed; a chase ensued, sometimes for over an hour before he could be captured, repacked, and again placed on the trail.

After performing a most arduous and difficult day's journey of fifteen miles, over continuous ranges of snowy ridges, we discovered that the mule on which was packed the bales of red cloth and blankets, intended for trading with the Indians, was missing. The muleteers did not remember to have seen him during the day. The animal was well trained, and was considered as one of the most willing and docile mules in the lot. Two men were sent back to look for him; it was easy to see if he had left the track, for the snow was unbroken, except on the trail make by our own party.

The men not returning in good time, we became

alarmed; they, however, made their appearance late in the night, with our lost mule; he was found standing, exactly in the same place where he was packed, behind a tree.

When the animals were driven out of camp, he was partly out of sight, escaped the vigilance of the men, and remained stationary, until our men found him in the evening; a lapse of at least twelve hours.

This incident is related to show the value of Mexican mules as faithful beasts of burthen, on which a great deal of dependence can always be placed. I consider them much preferable for travelling over the plains and mountains; they possess greater powers of endurance under privations. A mule will thrive on provender, that would starve a horse. If a mule gives out from exhaustion; with a day's rest, and a good meal, he will start on his journey, and appear as fresh as he ever was; but if a horse once stops and gives up, it is over with him, he is never fit for travel again. I suppose the noble and willing spirit of the horse, incites him to work until he is incapable of further exertion.

COL. FREMONT'S HORSE.

Col. Fremont started from Westport with a splendid dark bay horse; he was the pride of the party; he was always at the head of the cavalcade, and would sometimes look around, as it were disdainfully on his more humble companions. He felt his breeding, and I have no doubt, knew that he was carrying a gallant officer on his back. The Indians on the plains would have stolen him, and the Indians of the mountains would have given half-a-dozen mustangs for him. Mr. Palmer's horse gave out, and, was consequently on foot. We had at this time, the Doctor's mule, which we called "the Doctor," after he left us at Bent's Fort—running loose, as a spare animal, to carry the scientific apparatus. Col. Fremont the next day, rode the Doctor, and mounted Mr. W. H. Palmer on his own horse, which he continued to ride for ten days, until he was so exhausted for want of food, that he stopped on the road, and could not be brought into camp. Mr. Palmer came into camp on foot, and told Col. Fremont that his horse was left about five miles on the road, that it was impossible to bring him in.

I shortly afterwards heard Col. Fremont give orders to the Delaware camp, to send out a couple of men to find the horse and shoot it through the head. He had too much affection for the noble animal, to allow him to become a living sacrifice to the voracious wolves. The finer feelings of his heart seemed to govern all his actions, as well towards man as beast.

When it became necessary to slaughter our animals for food, I refrained from eating it in the vain hope of killing game, until exhausted nature demanded recuperation. I then partook of the strange and forbidden food with much hesitation, and only in small quantities. The taste of young fat horse meat is sweet and nutty, and could scarcely be distinguished from young beef, while that of the animal after it is almost starved to death, is without any flavor; you know you are eating flesh, but it contains no juices—it serves to sustain life, it contains but little nutritive matter, and one grows poor and emaciated, while living on it alone. Mule meat can hardly be distinguished from horse meat, I never could tell the difference. During one of the intervals when we were, from our own imprudence, entirely without food, a Delaware killed a cayotte, brought it into camp, and divided it equally between our messes—my share remained untouched. I had fasted 24 hours, and preferred to remain as many hours longer rather than partake of it. The habits of the horse and mule are clean; their food consists of grass and grain; but I was satisfied that my body could receive no benefit from eating the flesh of an animal that lived on carrion. Those who did partake of it were all taken with cramps and vomiting.

An old raven that had been hovering around us for several days, "to gather the crumbs from the rich man's table," paid at last the penalty of his temerity by receiving a rifle ball through his head. One of the men picked the feathers from its fleshless body and threw the carcass on the ground before us. It lay there undevoured when we left camp. I have no doubt it subsequently gave employment to a brother raven.

A RAIN STORM.

At the close of a long day's journey we descended into a fertile, although unknown, narrow valley, covered with dense forests of trees; a clear stream of water glided over its rocky bed, in the centre, and immense high sandstone mountains enclosed us; we chose a camp near the entrance of the valley, having deviated from our course, which was over the table land 500 feet above us, to obtain wood and water.

It is not at all improbable that our party were the first white men that ever penetrated into it—it was in reality a primeval forest. Our feet sank deep into the bed of dead leaves, huge trunks of trees in all stages of decay lay strewed around us, while trees of many kinds, were waving aloft their majestic limbs covered with spring foliage, shading our pathway. On the margin of the river grass of good quality grew in abundance, which afforded a delightful meal for our wearied animals. Although there was no snow visible around us, still the weather was cold and raw, the heavens were filled with floating clouds which seemed to increase as the night advanced. Large camp fires were soon burning, and another of our faithful horses was shot for food.

Selecting as I thought a comfortable place for my sleeping apartment, I made up my bed, placing as usual my India-rubber blanket on the decayed leaves. After supper I laid myself down to rest my exhausted body.

I had been on foot all day, travelling over a rugged country of volcanic formation, with an apology for moccasins on my lacerated and painful feet. I slept soundly until twelve o'clock, when I felt the cold water insinuating itself between my clothes and body. I uncovered my head, over which I had my robe and blankets, to find it raining fast and steadily. In an hour, I found myself laying in water nearly a foot deep. I could not escape from my present situation. Wrapping my India-rubber closely around me, I remained perfectly passive, submitting to the violence of the heaviest and most drenching rain-storm I experienced on the whole journey.

Darkness reigned supreme. Our camp-fires were extinguished, and but for the occasional ejaculations of our men, only the furious raging of the tempest, and the roar of the streams that came bounding in torrents from the table land above, could be heard. My blank ets and robe became saturated with water, while my clothes were wet to the skin.

I had ample time to reflect on my position; but while I experienced much personal inconvenience from the storm, the parched earth, over which we had travelled miles without a drop of water, received fresh sustenance from the refreshing shower. The dry and withered grass on our forward path, would be replaced by young tender shoots for our animals to sustain themselves. It is a happy thing for us that futurity is impenetrable, else my fond and fragile friends at home would endure more anguish than they do now, in their ignorance of the situation their husband and son is placed in.

Morning at last dawned, and with it appeared the sun, dissipating the clouds. Our camp equipage was all soaked. The daguerreotype apparatus was unhurt; my careful precaution always securing it against snow or rain. My polishing buffs I used the next day, when we ascended the mountain; I found them perfectly dry, and worked successfully with them. We remained late in camp the next morning to dry our blankets, etc. This was the first and only real storm of rain, we encountered in a six months' journey.

CHAPTER XIX.

Crippled Condition of the Party—Mr. Oliver Fuller—Mr. Egloffstien—Mr. Fuller gives out—His Inability to Proceed—Mr. Egloffstien and the Author continue on to Camp for Assistance—Col. Fremont sends Frank Dixon after him—Sorrow of the Camp—Mr. Fuller's Non-Appearance—Delawares sent out to Bring the Men in—Return of Frank Almost Frozen—Restoration of Mr. Fuller—Joy of the Men—Serious Thoughts—The Author Prepared to Remain on the Road—His Miraculous Escape.

Mr. Egloffstien, Mr. Fuller, and myself were generally at the end of the train, our scientific duties requiring us to stop frequently on the road. Mr. Fuller had been on foot several days before any of the rest of the party, his horse having been the first to give out. On this occasion, we started out of camp together. We were all suffering from the privations we had endured, and, of the three, I was considered the worst off. One of my feet became sore, from walking on the flinty mountains with thin moccasins, and I was very lame in consequence. Mr. Fuller's feet were nearly wholly exposed. The last pair of moccasins I had, I gave him a week before; now his toes were out, and he walked with great difficulty over the snow. He never complained when we started in the morning, and I was surprised when he told me he had " given out."

"Nonsense, man," I said; "let us rest awhile, and we will gather fresh strength." We did so, and at every ten steps he had to stop, until he told us that he could go no further.

Mr. Fuller was the strongest and largest man in camp when we left Westport, and appeared much better able to bear the hardships of the journey than any man in it. I was the weakest, and thought ten days before that I would have given out, yet I live to write this history of his sufferings and death, and to pay this tribute to his memory.

The main body of the camp had preceded us, and they were at least four miles a-head. Both Mr. Egloffstien and myself offered our personal assistance; Mr. Fuller leaned upon us, but could not drag one foot after the other—his legs suddenly becoming paralyzed. When we realized his condition, we determined to remain with him; to this he decidedly objected—"Go on to camp," said he, "and if possible, send me assistance. You can do me no good by remaining, for if you do not reach camp before night, we shall all freeze to death."

He luckily had strapped to his back his blue blankets, which we carefully wrapped around him. In vain we hunted for an old bush or something with which to make a fire—nothing but one vast wilderness of snow was visible. Bidding him an affectionate farewell, and promising to return, we told him not to move off the trail, and to keep awake if possible.

Limping forward, Egloffstien and myself resumed our travel; the sun had passed the meridian, and dark clouds overhung us. The night advanced apace, and with it an increase of cold. We stopped often on the road, and with difficulty ascended a high hill, over which the trail led; from its summit I hoped to see our camp-fires; my vision was strained to the utmost, but no friendly smoke greeted my longing eyes. The trail lost itself in the dim distance, and a long and weary travel was before us. Nothing daunted, and inspired by the hope of being able to render succor to our friend, we descended the mountain and followed the trail.

It now commenced to snow. We travelled in this manner ten long hours, until we came upon the camp.

Mr. Egloffstien and self both informed Col. Fremont of the circumstance, and we were told that it was impossible to send for Mr. Fuller.

Overcome with sorrow and disappointment, I fell weeping to the ground. In my zeal and anxiety to give assistance to my friend, I never for a moment thought in what manner it was to be rendered. I had forgotten that our few remaining animals were absolutely necessary to carry the baggage and scientific apparatus of the expedition, and that, with a furiously-driving snow-storm, it was almost folly to attempt to find the trail.

While we were speaking at our scanty fire of the unfortunate fate of our comrade, Col. Fremont came out of his lodge, and gave orders that the two best animals in camp should be prepared, together with some cooked horse-meat. He sent them with Frank Dixon, a Mexican, back on the trail, to find Mr. Fuller. We supposed him to have been at least five miles from camp.

There was not a dry eye in camp that whole night. We sat up anxiously awaiting the appearance of Mr. Fuller. Col. Fremont frequently inquired of the guard if Mr. Fuller had come in?

Day dawned, and cold and cheerless was the prospect. There being no signs of our friend, Col. Fremont remarked that it was just what he expected.

Col. Fremont had allowed his humanity to overcome his better judgment.

At daylight, Col. Fremont sent out three Delawares to find the missing men; about ten o'clock one of them returned with Frank Dixon, and the mules; Frank had lost the trail, he became bewildered in the storm, and sank down in the snow, holding on to the mules. He was badly frozen, and became weaker every day until he got to the settlements. Towards night, the two Delawares supporting Mr. Fuller, were seen approaching; he was found by the Delawares awake, but almost senseless from cold and starvation; he was hailed with joy by our whole camp. Col. Fremont as well as the rest of us, rendered him all the assistance in our power; I poured out the last drop of my alcohol, which I mixed with a little water, and administered it to him. His feet were frozen black to his ancles; if he had lived to reach the settlements, it is probable he would have had to suffer amputation of both feet.

Situated as we were, in the midst of mountains of snow, enervated by starvation and disease, without animals to carry us, and a long uncertain distance to travel over an unexplored country; could any blame be attached to a commander of an expedition, if he were to refuse to send back for a disabled man? I say, no, none whatever. Twenty-seven of our animals had been killed for food, and the rest were much reduced, and without provender of any kind in view. If this event had occurred six days later, there would have been no animal strong enough to carry Mr. Fuller into camp.

But suppose he had been disabled while in camp, and unable to proceed, could blame attach to his comrades if he were deserted, and left to die alone? This frightful situation was nearly realized on several occasions. I again answer, no, not any—the safety of the whole party demanded their immediate extrication from the dangers which surrounded them; every hour, every minute, in these mountains of snow, but increased their perils; on foot, with almost inaccessible rugged mountains of snow to overcome, with no prospects of food except what our remaining animals might afford—to stop, or remain an indefinite time with a disabled comrade, was certain death to the whole party, without benefiting him; his companions being so weak, that they could not carry him along. I made up my mind on one occasion, not to leave camp, my exhausted condition reminded me of the great difficulty and bodily pain which I endured, to reach camp the night before. I was fully prepared to remain by myself, and await my fate. I probably should have done so, but for the fond links which bound me to life, exercising a magic influence which inspired me with fresh courage, and determination. If such had been the case, might not my friends, in the excess of their grief have exclaimed, "Alas! for my poor son, who was left by his companions to perish in the mountains of snow." It would have been difficult to have persuaded my old parents, of the utter impossibility of preventing it. They would have attached cruelty, and neglect, to the whole party, and laid their son's death at the door of their leader.

How is it in war, when the superior force of the enemy demands an immediate retreat by the opposing army, without permitting time to carry the wounded off the field? How is it with a man who falls overboard during a storm, when imminent peril to the vessel and crew would follow an attempt to rescue him? The life of one must be sacrificed for the safety of the whole.

CHAPTER XX.

Author nearly gives Out—Family Portraits—Fresh Courage—Dangerous Situation- Lonely Journey—Darkness—Snow Storm—Arrival at Camp—" Col. Fremont's Tent —Interview with Col. Fremont—" Cache "—Men on Foot—Daguerreotype Apparatus buried in the Snow—Sperm Candles—Men Mounted on Baggage Animals— Seveir River Beaver Dams—Modus Operandi of killing Horses for Food—Entrail Soup—Hide and Bones Roasted—Influence of Privation on Human Passions.

AFTER we crossed the Green River, the whole party were on foot. The continued absence of nutritious food made us weaker every day. One of my feet was badly frozen, and I walked with much pain and great difficulty; on this occasion my lameness increased to such a degree, that I was the last man on the trail, and my energy and firmness almost deserted me. Alone, disabled, with no possibility of assistance from mortal man, I felt that my last hour had come; I was at the top of a mountain of snow, with not a tree to be seen for miles. Night approached, and I looked in vain in the direction our party had proceeded, for smoke or some indication that our camp was near. Naught but a desert waste of eternal snow met my anxious gaze— faint and almost exhausted, I sat down on the snow-bank, my feet resting in the footsteps of those who had gone before me. I removed from my pocket the miniatures of my wife and children, to take a last look at them. Their dear smiling faces awakened fresh energy, I had still something to live for, my death would bring

AUTHOR'S ENERVATED CONDITION. 123

heavy sorrow and grief to those who looked to me alone for support; I determined to try and get to camp, I dared not rest my fatigued body, for to rest was to sleep, and sleep was that eternal repose which wakes only in another world. Offering up a silent prayer, I prepared to proceed. I examined my gun and pistols, so as to be prepared if attacked by wolves or Indians, and resumed my lonely and desolate journey. As the night came on, the cold increased; and a fearful snow storm blew directly in my face, almost blinding me. Bracing myself as firmly as I could against the blast, I followed the deep trail in the snow, and came into camp about ten o'clock at night. It requires a personal experience to appreciate the intense mental suffering which I endured that night; it is deeply engraven with bitter anguish on my heart, and not even time can obliterate it.

Col. Fremont was at the camp fire awaiting my arrival. He said he knew I was badly off, but felt certain I would come in, although he did not expect me for an hour.

My haggard appearance sufficiently indicated what I suffered. As I stood by the fire warming my frozen limbs, Col. Fremont put out his hand and touched my breast, giving me a slight push; I immediately threw back my foot to keep myself from falling. Col. Fremont laughed at me and remarked that I had not "half given out," any man who could act as I did on the occasion, was good for many more miles of travel. He went into his tent, and after my supper of horse soup, he sent for me, and then told me why he played this little joke on me; it was to prevent my telling my sufferings to the men; he saw I had a great deal to say, and that no good would result from my communicating it. He reviewed our situation, and the enervated condition of the men, our future prospects of getting into settlements, and the necessity there was for mutual encouragement, instead of vain regrets, and despondency; the difficulties were to be met, and it depended on ourselves, whether we should return to our families, or perish on the mountains; he bade me good night, telling me that in the morning he would endeavor to make some arrangements to mount the men.

The next day, he called the men together and told them that he had determined to "cache" all the superfluous baggage of the camp, and mount the men on the baggage animals, as a last resource. Nothing was to be retained but the actual clothing necessary to protect us from the inclemency of the weather.

A place was prepared in the snow, our large buffalo lodge laid out, and all the pack saddles, bales of cloth and blankets, the travelling bags, and extra clothes of the men, my daguerreotype boxes, containing besides, several valuable scientific intruments, and everything that could possibly be spared, together with the surplus gunpowder and lead, were placed in it, and carefully covered up with snow, and then quantities of brush to protect it from the Indians. I previously took out six sperm candles from my boxes, and gave them to Lee, the Colonel's servant, in charge; they were subsequently found most useful. A main station was made at this place, so as to be able to find it if occasion demanded that we should send for them.

The men now were all mounted; a large mule was allotted to me, and we again started, rejoicing in having animals to carry us. After this, every horse or mule that gave out, placed a man on foot without the possibility of procuring others, and it was necessary in consequence of the absence of grass, to allow the mules to travel as light as possible; we therefore relieved them frequently by walking as much as we were able.

BEAVER DAMS ON THE SEVEIR RIVER.

When we got to the crossing of the Seveir River, I was almost certain I was within the precincts of civilization. I saw numberless large trees cut down near the roots, appearing to have been hewn with an axe; some of them laid directly across the river; in one place there were three trees lying parallel with each other, evidently intended, I supposed, as a bridge across it; at this spot, the stream was not more than thirty feet wide; no other indication of civilization being around us, I supposed we occupied an old camping ground of Indians. I was doomed to disappointment again; the beavers had constructed the dams, and cut down the trees, and not until I had closely inspected the work, could I believe that they were not the work of men.

MANNER OF DIVIDING THE HORSE MEAT.

When an animal gave out, he was shot down by the Indians, who immediately cut his throat, and saved the blood in our camp kettle. (The blood I never partook of.) The animal was divided into twenty-two parts as follows:—Two for Col. Fremont and Lee, his cook; ten for the Delaware camp, and ten for ours. Col. Fremont hitherto had messed with his officers; at this time he requested that we should excuse him, as it gave him pain, and called to mind the horrible scenes which had been enacted during his last expedition—he could not see his officers obliged to partake of such disgusting food.

The rule adopted was, that one animal should serve for six meals for the whole party.

If one gave out in the meantime, of course it was an exception, but otherwise on no consideration was an animal to be slaughtered, for every one that was killed placed one man on foot, and limited our chance of escape from our present situation.

If the men chose to eat up their six meals in one day, they would have to go without until the time arrived for killing another.

It frequently happened that the white camp was without food from twenty-four to thirty-six hours, while Col. Fremont and the Delawares always had a meal.

The latter religiously abstained from encroaching on the portion allotted for another meal while many men of our camp, I may say all of them, not content with their daily portion, would, to satisfy the cravings of hunger, surreptitiously purloin from the pile of meat at different times, sundry pieces thus depriving themselves of each other's allowance.

The entrails of the horse were well shaken (for we had no water to wash them in) and boiled with snow, producing a highly flavored soup, peculiar to itself, and readily distinguished from the various preparations of the celebrated "Ude" of gastronomic memory. The hide was roasted so as to burn the hair and make it crisp, the hoofs and shins were disposed of by regular rotation.

Our work was never done. When we got to camp

all the men off duty, were dispatched to gather firewood to burn during the night. One might be seen with a decayed trunk on his shoulder, while a half dozen others were using their combined efforts to bring into camp some dried tree.

Col. Fremont at times joined the men in this duty—when it was peculiarly difficult in procuring the necessary material to prevent us from freezing while we were in camp.

One night we camped without wood, the country around was a waste of snow; we laid down in our blankets, and slept contentedly till morning, and re-commenced our journey without any breakfast.

I have been awakened to go on "guard" in the morning watch, when, looking around me, my companions appeared like so many graves, covered with from eight to ten inches of snow.

Some of our animals would eat the snow, others would not. To keep them alive we had to melt snow in camp kettles and give it them to drink, which process was attended with much fatigue and trouble.

We lived on horse meat fifty days. The passions of the men were so disturbed by their privations, that they were not satisfied with the cook's division of the hide; but one man turned his back, while another asked him who was to have this piece, and that, and so on, until all was divided, and the same process was gone through with in the sharing of the delectable horse soup.

CHAPTER XXI.

Unsuccessful Attempt to Force a Passage in the Mountains—Delawares sent out to Explore—Their Return—Col. Fremont, Capt. Wolff, and Solomon in Council—Unfavorable Report of Capt. Wolff—Col. Fremont's Determination—Astronomical Observations at Midnight—Col. Fremont's Correctness and Skill Illustrated—Tremendous Mountains of Snow—Successful Ascent on Foot, without Shoes or Moccasins—Tribute to the Genius of Fremont—Col. Fremont's Lodge at Meal-Time—Mr. Oliver Fuller's Death—Sorrow of his Companions—His Last Hours—His Virtues—Indian Camp—Arrival at Parowan—Burial of Mr. Fuller—Author's Physical Condition—Mormon Sympathies—Mr. Heap and his Wives—Mormon Hospitality.

Four days before we entered the Little Salt Lake Valley, we were surrounded by very deep snows; but as it was necessary to proceed, the whole party started, to penetrate through what appeared to be a pass, on the Warsatch Mountains. The opening to this depression was favorable, and we continued our journey, until the mountains seemed to close around us, the snow in the cañon got deeper, and further progress on our present course was impossible.

It was during this night, while encamped in this desolate spot, that Col. Fremont called a council of Capt. Wolff and Solomon of the Delawares—they had been sent by Col. Fremont to survey the cañon, and surrounding mountains, to see if a passage could be forced. On their return, this council was held; Capt. Wolff reported it impossible to proceed, as the animals sank over their heads in snow, and he could see no passage out. The mountains which intercepted our path, were covered with snow four feet deep. The ascent bore an angle of forty-five degrees, and was at least one thousand feet from base to summit. Over this, Captain Wolff said it was also impossible to go. "That is not the point," replied Col. Fremont, "we must cross, the question is, which is most practicable—and how we can do it."

I was acting as assistant astronomer at this time. After the council, Col. Fremont told me there would be an occultation that night, and he wanted me to assist in making observations. I selected a level spot on the snow, and prepared the artificial horizon. The thermometer indicated a very great degree of cold; and standing almost up to our middle in snow, Col. Fremont remained for hours making observations, first with one star, then with another, until the occultation took place. Our lantern was illuminated with a piece of sperm candle, which I saved from my pandora box, before we buried it; of my six sperm candles this was the last one. I take some praise to myself for providing some articles which were found most necessary. These candles, for instance, I produced when they were most required, and Col. Fremont little thought where they were procured.

The next morning, Col. Fremont told me that Parowan, a small settlement of Mormons, forty rods square, in the Little Salt Lake Valley, was distant so many miles in a certain direction, immediately over this great mountain of snow; that in three days he hoped to be in the settlement, and that he intended to go over the mountain, at all hazards.

We commenced the ascent of this tremendous mountain, covered as it were, with an icy pall of death, Col. Fremont leading and breaking a path; the ascent was so steep and difficult, that it was impossible to keep on our animals; consequently, we had to lead them, and travel on foot—each man placed his foot in the tracks of the one that preceded him; the snow was up to the bellies of the animals. In this manner, alternately toiling and resting, we reached the summit, over which our Delawares, who were accustomed to mountain travel, would not of themselves have ventured. When I surveyed the distance, I saw nothing but continued ranges of mountains of everlasting snow, and for the first time, my heart failed me—not that I had lost confidence in our noble leader, but that I felt myself physically unable to overcome the difficulties which appeared before me, and Capt. Wolff himself told me, that he did not think we could force a passage. We none of us had shoes, boots it was impossible to wear. Some of the men had raw hide strapped round their feet, while others were half covered with worn out stockings and moccasins; Col. Fremont's moccasins were worn out, and he was no better off than any of us.

After we were all rested, Col. Fremont took out his pocket compass, and pointing with his hand in a certain direction, commenced the descent. I could see no mode of extrication, but silently followed the party, winding round the base of one hill, over the side of another, through defiles, and, to all appearance, impassable cañons, until the mountains, which were perfectly bare of vegetation, gradually became interspersed with trees. Every half hour, a new snow scape presented itself, and as we overcame each separate mountain, the trees increased in number.

By noon, we were in a defile of the mountains, through which was a dry bed of a creek. We followed its winding course, and camped at about two o'clock in a valley, with plenty of grass. Deer tracks were visible over the snow, which gave fresh life to the men. The Delawares sallied out to find some. Col. Fremont promised them, as an incentive to renewed exertions, that he would present to the successful hunter, who brought in a deer, a superior rifle.

They were out several hours, and Weluchas was seen approaching, with a fine buck across his saddle.

He received his reward, and we again participated in a dish of wholesome food.

We had now triumphantly overcome the immense mountain, which I do not believe human foot, whether civilized or Indian, had ever before attempted, from its inaccessibility; and on the very day and hour previously indicated by Col. Fremont, he conducted us to the small settlement of Parowan, in Little Salt Lake Valley, which could not be distinguished two miles off, thus proving himself a most correct astronomer and geometrician.

Here was no chance work—no guessing—for a deviation of one mile, either way, from the true course, would have plunged the whole party into certain destruction. An island at sea may be seen for forty miles; a navigator makes his calculations, and sails in the direction of the land, which oftentimes extends many miles; when he sees the land, he directs his course to that portion of it where he is bound; he may have been fifty miles out of his way, but the well-known land being visible from a great distance, he changes his course until he arrives safely in port.

Not so with a winter travel over trackless mountains of eternal snow, across a continent of such immense limits, suffering the privations of cold and hunger, and enervated by disease.

It seems as if Col. Fremont had been endowed with supernatural powers of vision, and that he penetrated with his keen and powerful eye through the limits of space, and saw the goal to which all his powers had been concentrated to reach. It was a feat of scientific correctness, probably without comparison in the records of the past. His firmness of purpose, determination of character, and confidence in his own powers, exercised under such extraordinary circumstances, alone enabled him, successfully, to combat the combination of untoward and unforeseen difficulties which surrounded him, and momentarily threatened the annihilation of his whole party.

It is worthy of remark, and goes to show the difference between a person "to the manor born," and one who has "acquired it by purchase." That in all the varied scenes of vicissitude, of suffering and excitement, from various causes, during a voyage when the natural character of a man is sure to be developed, Col. Fremont never forgot he was a gentleman; not an oath, no boisterous ebullutions of temper, when, heaven knows, he had enough to excite it, from the continued blunders of the men. Calmly and collectedly, he gave his orders, and they were invariably fulfilled to the utmost of the men's abilities. To the minds of some men, excited by starvation and cold, the request of an officer is often misconstrued into a command, and resistance follows as a natural consequence; but in no instance was a slight request of his received with anything but the promptest obedience. He never wished his officers or men to undertake duties which he did not readily share. When we were reduced to rations of dried horse meat, and he took his scanty meal by himself, he was, I am sure, actuated by the desire to allow his companions free speech, during meal time; any animadversion on the abject manner in which we were constrained to live would, no doubt, have vibrated on his sensitive feelings, and to prevent the occurrence of such a thing, he, as it were, banished himself to the loneliness of his own lodge.

Col. Fremont's lodge, at meal time, when we had good, wholesome buffalo and deer meat presented quite a picturesque appearance. A fire was always burning in the centre; around it cedar bushes were strewn on which buffalo robes were placed. Sitting around, all of us on our hams, cross-legged, with our tin plates and cups at each side of us, we awaited patiently the entrance of our several courses; first came the camp kettle, with buffalo soup, thickened with meat-biscuit, our respective tin plates were filled and replenished as often as required. Then came the roast or fry, and sometimes both; the roast was served on sticks, one end of which was stuck in the ground, from it we each in rotation cut off a piece. Then the fried venison. In those days we lived well, and I always looked forward to this social gathering, as the happiest and most intellectually spent hour during the day. Col. Fremont would often entertain us with his adventures on different expeditions; and we each tried to make ourselves agreeable.

Although on the mountains, and away from civilization, Col. Fremont's lodge was sacred from all and every thing that was immodest, light or trivial; each and all of us entertained the highest regard for him. The greatest etiquette and deference were always paid to him, although he never ostensibly required it. Yet his reserved and unexceptionable deportment, demanded from us the same respect with which we were always treated, and which we ever took pleasure in reciprocating.

MR. FULLER'S DEATH.

The death of Mr Fuller filled our camp with deep gloom; almost at the very hour he passed away, succor was at hand. Our party was met by some Utah Indians, under the chieftainship of Ammon, a brother of the celebrated Wakara, (anglicized Walker) who conducted us into the camp on Red Creek Cañon. At this spot our camp was informed by Mr. Egloffstien, that our companion in joy and in sorrow, was left to sleep his last sleep on the snows. The announcement took some of us by surprise, although I was prepared for his death at any moment. I assisted him on his mule that morning, and roasted the prickles from some cactus leaves, which we dug from the snow, for his breakfast; he told me that he was sure he would not survive, and did not want to leave camp.

A journey like the one we had passed through, was calculated to expose the thorough character of individuals; if there were any imperfections, they were sure to be developed. My friend, Oliver Fuller, passed through the trials of that ordeal victoriously. No vice. or evil propensity made any part of his character. His disposition was mild and amiable, and generous to a fault. Slow to take offence, yet firm and courageous as a lion; he bore his trials without a murmur, and performed his duties as assistant astronomer and engineer to the hour he was stricken down. After he was unable to walk, he received the assistance of every man in camp.

His companions who were suffering dreadfully, though not to such an imminent degree, voluntarily deprived themselves of a portion of their small rations of horse meat to increase his meal, as he seemed to require more sustenance than the rest of us. His death was deeply regretted.

Not having any instruments by which a grave could be dug in the frozen ground, Col. Fremont awaited his arrival at Parowan, from which place he sent out several men to perform the last sad duties to our lamented friend.

I was riding side by side with Egloffstien after Mr. Fuller's death, sad and dejected. Turning my eyes on the waste of snow before me, I remarked to my companion that I thought we had struck a travelled road. He shook his head despondingly, replying "that the marks I observed, were the trails from Col. Fremont's lodge poles." Feeling satisfied that I saw certain indications, I stopped my mule, and with very great difficulty alighted, and thrust my hand into the snow, when to my great delight I distinctly felt the ruts caused by wagon wheels. I was then perfectly staisfied that we were "saved!" The great revulsion of feeling from intense despair to a reasonable hope, is impossible to be described; from that moment, however, my strength perceptibly left me, and I felt myself gradually breaking up. The nearer I approached the settlement, the less energy I had at my command; and I felt so totally incapable of continuing, that I told Col. Fremont, half an hour before we reached Parowan, that he would have to leave me there; when I was actually in the town, and surrounded with white men, women and children, paroxysms of tears followed each other, and I fell down on the snow perfectly overcome.

I was conducted by a Mr. Heap to his dwelling, where I was treated hospitably. I was mistaken for an Indian by the people of Parowan. My hair was long, and had not known a comb for a month, my face was unwashed, and ground in with the collected dirt of a similar period. Emaciated to a degree, my eyes sunken, and clothes all torn into tatters from hunting our animals through the brush. My hands were in a dreadful state; my fingers were frost-bitten, and split at every joint; and suffering at the same time from diarrhœa, and symptoms of scurvy, which broke out on me at Salt Lake City afterwards. I was in a situation truly to be pitied, and I do not wonder that the sympathies of the Mormons were excited in our favor, for my personal appearance being but a reflection of the whole party, we were indeed legitimate subjects for the exercise of the finer feelings of nature. When I entered Mr. Heap's house I saw three beautiful children. I covered my eyes and wept for joy to think I might yet be restored to embrace my own.

During the day I submitted to the operation of having my face and hands washed, and my hair cut and combed. Our combs might have been lost, and this would account for the condition of our hair, but how about the dirty faces? Alas, we had no water, nothing but frozen snow; and although we laved our faces with it, we had no towels to wipe with, and the dirt dried in.

Mr. Heap was the first Mormon I ever spoke to, and although I had heard and read of them, I never contemplated realizing the fact that I would have occasion to be indebted to Mormons for much kindness and attention, and be thrown entirely among them for months.

It was hinted to me that Mr. Heap had two wives; I saw two matrons in his house, both performing to interesting infants the duties of maternity; but I could hardly realize the fact that two wives could be reconciled to live together in one house. I asked Mr. Heap if both these ladies were his wives, he told me they were. On conversing with them subsequently, I discovered that they were sisters, and that there originally were three sets of children; one mother was deceased, and she was also a sister. Mr. Heap had married three sisters, and there were living children from them all. I thought of that command in the bible,—"Thou shalt not take a wife's sister, to vex her." But it was no business of mine to discuss theology or morality with them—they thought it right.

These two females performed all the duties which devolve on a country home. One of them milked the cow, churned the butter, and baked the bread; while the other cared for the children, attended to the making, washing, and ironing of the clothes. Mr. Heap was an Englishman, and his wives were also natives of London. Mr. Heap was a shoemaker by trade, and a preacher by divine inspiration. Mammon was the god he worshipped, for he gave away nothing without an equivalent—not even a piece of old cloth to line a pair of moccasins with. His wives differed from him in this respect, daily they furnished " Shirt-cup," the " Utah,' with everything edible, for numbers of miserable Indians who surrounded their door. The eldest in particular, was a kind-hearted woman; they all, however, showed me as much attention as they could afford, for one dollar and fifty cents a day, which amount Col. Fremont paid for my board while with them, a period of fourteen days.

CHAPTER XXII.

Sojourn at Parowan—Colonel Fremont refits his Expedition—Illness of the Author—His Inabillity to Proceed—Takes Leave of Col. Fremont—Mr. Egloffstien and the Author leave to go to Great Salt Lake City in a Wagon—Col. Fremont's Departure—Mormons for Conference—Arrival at Salt Lake City—Massacre of Capt. Gunnison—Interview with Lieut. Beckwith—Mr. Egloffstien appointed Topographical Engineer—Painting Materials—Kinkead and Livingston—Brigham Young—Governor's Residence—Apology for Mormonism among the Masses—Their previous Ignorance of the Practice of Polygamy.

I REMAINED from the 8th to the 21st February at Parowan. I was very ill during the whole time; I was so much enervated by diarrhœa, that my physician advised me not to accompany the expedition; the exertion of riding on horseback would have completely prostrated me, my digestive organs were so much weakened, and impaired, by the irregular living on horse meat, without salt or vegetables, that I was fearful that I should never recover. Col. Fremont was very anxious for me to continue, but yielded to the necessity of my remaining; he supplied me with means to reach home, and on the same day he bade me farewell, to continue his journey over the Sierra Nevada, I left for great Salt Lake City, in a wagon belonging to one of a large company of Mormons, who were on their way to "Conference." I was so weak, that I had to be lifted in and out like a child. To the kind attentions of Mr. Henry Lunt, President of Cedar City, Coal Creek, and his lady, I was indebted for some necessaries, viz.—sugar, tea and coffee, which it was im possible to purchase; they also offered me the use of their wagon, which was better adapted to an invalid, than the one I occupied. Mr. Egloffstien also accompanied me; his physical condition being similar to my own, he could not continue with Col. Fremont; he successfully managed, notwithstanding his illness, to make topographical notes all the way to Great Salt Lake City, a distance of three hundred miles, which we accomplished in ten days, passing through all the different Mormon settlements on the road, particulars of which I shall give in my journal, from Salt Lake City. We arrived at Great Salt Lake City on the night of the 1st of March 1854, and took lodgings at Blair's hotel; in the morning I learned that Lieut. Beckwith and Captain Morris, with the remnant of Captain Gunnison's expedition, were hyber nating in the city. I called on Lieut. Beckwith, who invited me and my friend to mess at their table, at E. T. Benson's, one of the Mormon apostles, which I gladly accepted, and that night I found myself once more associating with intelligent gentlemen. The arrival of my friend Egloffstein, proved very timely; the massacre of the lamented Captain Gunnison and his officers, deprived Lieut. Beckwith of the services of their topographical engineer, to which situation Mr. Egloffstein was immediately appointed, and Lieut. Beckwith generously invited me to accompany the expedition, free of any expense, which I respectfully declined, as I intended to reach California by the Southern route, over the trail of Colonel Fremont, in 1843. To the kindness of Lieut. Beckwith I was also indebted for a supply of painting materials, which I could not have procured elsewhere, and by the use of which, I was enabled to successfully prosecute my profession, during my residence in that city.

Messrs. Kincaid and Livingston, cashed Col. Fremont's bills on California, without any discount, and contributed many luxuries which were not on sale, and I feel deeply grateful to them for their disinterested friendship. After I was comfortably settled, I called on Governor Young, and was received by him with marked attention. He tendered me the use of all his philosophical instruments and access to a large and valuable library.

The court-house of the city of the Great Salt Lake lies in 40° 45′ 44″ N. Lat. 111° 26′ 34″ W. Longitude, and the city covers an area of four square miles, it is laid out at right angles. The principal business streets run due north and south, a delicious stream of water flows through the centre of the city, this is subdivided into murmuring rivulets on either side of all the streets. The water coming directly from the mountains, is always pure and fresh, affording this most useful element in any quantity, and within reach of every one, besides creating a healthful influence in the city. Cotton-wood trees grow on the main stream, and saplings had just been planted while I was there, on the sides of the streets. Most of the dwelling-houses are built a little distance from the side-walk, and to each dwelling is appropriated an acre and a quarter of ground, for gardening purposes.

Salt Lake Valley runs east and west, and the city is immediately at the base of a high range of mountains. An adobe wall, twelve feet high, six feet at the base, tapering upwards to 2½ feet, entirely surrounds the city, enclosing an immense area of ground for pasturage, etc. thus protecting the people and cattle from the aggressions of Indians. The Timpanagos mountains are near the city: "Emigration Cañon" is the gate (a low depression in the mountains) through which the great tide of emigration flows into the Valley of Great Salt Lake.

The River Jordan runs through the valley and empties into Great Salt Lake. The city is thirty miles from the Lake, and the valley is entirely surrounded with high mountains topped with snow, winter and summer.

The governor's residence, a large wooden building of sufficient capacity to contain his extensive family—nineteen wives and thirty-three children, was nearly finished. I made a daguerreotype view of it, and also a drawing.

The court house is a large square building, on the east side, opposite the Temple square.

The post office occupies the corner on the south side.

The Tabernacle, an unpretending one story building, occupies a portion of the Temple square.

The Temple is in course of building—the foundation is laid—and I was allowed to see the plan projected by a Mr. Angell, who by *inspiration* has succeeded in producing an exact model of the one used by the Melchizedek Priesthood, in older times.

The theatre, a well built modern building, is opposite to the governor's house on the north, and is the property of the church as are all the public buildings. I may say all the real estate in the valley is the property of the church, for proprietors have only an interest in property so long as they are members of the Mormon Church, and reside in the valley. The moment they leave or apostatize, they are obliged to abandon their property, and are precluded from selling it, or if they do give the bill of sale it is not valid—it is not tenable by the purchaser. This arrangement was proposed by the governor and council, at the conference which took place during my residence among them in 1854, and thousands of property holders subsequently deeded their houses and lands to the church, in perpetuity.

Under the operation of this law, nobody but Mormons can hold property in Great Salt Lake City. There are numbers of citizens who are not Mormons, who rent properties; but there is no property for sale—a most politic course on the part of the Mormons—for in case of a railroad being established between the two oceans, Great Salt Lake City must be the half way stopping place, and the city will be kept purified from taverns and grog shops at every corner of the street. Another city will have to be built some distance from them, for they have determined to keep themselves distinct from the vices of civilization. During a residence of ten weeks in Great Salt Lake City, and my observations in all their various settlements, amongst a homogeneous population of over seventy-five thousand inhabitants, it is worthy of record, that I never heard any obscene or improper language; never saw a man drunk; never had my attention called to the exhibition of vice of any sort. There are no gambling houses, grog shops, or buildings of ill fame, in all their settlements. They preach morality in their churches and from their stands, and what is as strange as it is true, the people practise it, and religiously believe their salvation depends on fulfilling the behests of the religion they have adopted.

The masses are sincere in their belief, if they are incredulous, and have been deceived by their leaders, the sin, if any, rests on them. I firmly believe the people to be honest, and imbued with true religious feelings,—

and when we take into consideration their general character previously, we cannot but believe in their sincerity. Nine-tenths of this vast population are the peasantry of Scotland, England and Wales, originally brought up with religious feelings at Protestant parish churches. I observed no Catholic proselytes. They have been induced to emigrate, by the offers of the Mormon missionaries to take them free of expense, to their land flowing with milk and honey, where, they are told, the Protestant Christian religion is inculcated in all its purity, and where a farm and house are bestowed gratuitously upon each family. Seduced by this independence from the state of poverty which surrounds them at home, they take advantage of the opportunity and are baptized into the faith of the "latter day saints," and it is only after their arrival in the Valley that the spiritual wife system is even mentioned to them. Thousands of families are now in Utah who are as much horrified at the name of polygamy, as the most carefully educated in the enlightened circles of Europe and America. More than two-thirds of this population (at least, this is the ratio of my experience) cannot read or write, and they place implicit faith in their leaders, who, in a pecuniary point of view, have fulfilled their promise; each and all of them are comfortably provided with land and tenements. The first year they, of course, suffer privations, until they build their houses and reap their crops, yet all their necessities in the meantime are provided for by the church, and in a social point of view, they are much happier than they could ever hope to have been at their native homes. From being tenants at will of an imperious and exacting landlord, they suddenly become land holders, in their own right—free

APOLOGY FOR MORMONISM. 145

men, living on free soil, under a free and enlightened government.

Their religious teachers of Mormonism, preach to them, as they call it, "Christianity in its purity." With their perfect right to imbibe new religious ideas, I have no wish to interfere, nor has any one. All religions are tolerated, or ought to be, in the United States, and I offer these remarks as an apology for the masses of honest men, many of whom have personally told me, that they were ignorant of the practice of polygamy before their arrival in the Valley, and surrounded as they are, by hostile tribes of Indians, and almost unsurmountable mountains of snow, they are precluded from returning home, but live among themselves, practicing as well as they know how, the strict principles of virtue and morality.

CHAPTER XXIII.

Governor Brigham Young—Author's Views on Polygamy—Baptismal Ceremony--Doctrines and Covenants.

I RECEIVED a good deal of marked attention from his excellency, Governor Young; he often called for me to take a drive in his carriage, and invited me to come and live with him, during the time I sojourned there. This invitation I refused, as I wished to be entirely independent to make observations. I told Brigham Young that I was making notes, with a view to publish them. He replied, " Only publish 'facts,' and you may publish as many as you please." I shall, in the succeeding chapters, give personal relations of facts, with alterations of names only, not wishing to bring the real actors before the public. I offer them to show up the *abuses* which a polygamous life must be subjected to, when human passions are allowed free scope, and not subject to laws, either social or moral. I hope to live to see a more wholesome feeling, in this respect, among the leaders of the Mormon Church. A continuation of their present practice must inevitably lead to confusion.

CEREMONY OF BAPTISM.

March 30th.—The weather is very cold, and snow lies on the ground to the depth of six inches.

CEREMONY OF BAPTISM.

A stream of living water, twelve feet wide, fresh from the mountains, runs along between the sidewalk and the road—the Temple Block. Seeing a crowd assembled, I approached the spot, and found twelve persons, some of whom had already undergone the ceremony of baptism, and others patiently awaiting. The first immersion I saw was of a lady about 18 years of age. The priest who officiated, was standing up to his waist in the stream, with his coat off, and his sleeves rolled up to his elbows. The lady was handed in, and I noticed the shock on her system which a sudden plunge into cold freezing water must naturally have produced. The baptizer placing one hand on her back, the other on her head, repeated the following words: "I am commissioned by Jesus Christ to baptize you, in the name of the Father, and Son, and of the Holy Ghost. Amen."

He then pushed her over on her back, allowing the water to cover her. She struggled to get out of the water, but her husband remarked that the whole of her head had not been submerged, and insisted that "his wife should be properly baptized." She was consequently dipped effectually a second time, and the poor woman finally made her escape, almost frozen.

The next subject was an old lady of seventy-five years. She tottered into the stream by the aid of her crutch, and underwent the same ceremony. Query: would persons submit to those extraordinary tests if they did not possess faith?

The third person was a young man of about twenty years, with a calm, placid countenance. He underwent the operation without flinching. His face was the impersonation of faith and purity. I should have liked to have painted him as a study for a "St. John." They went each on their respective ways, many of them, I dare say, with the seeds of consumption sown at this moment, fully determined to live a life of piety and virtue.

* * * * * *

The men, after baptism, are elders, and are empowered to perform the ceremony upon others. They wear an under-garment with distinctive marks upon it, in imitation of the Jews, "who all wear fringes on the borders of their garments, that they may look upon them and remember the commandments of the Lord to do them."—*Deuteronomy*.

There are two priesthoods in the Mormon Church: the Melchizedek and the Aaronic, including the Levitical. The office of an elder comes under the Melchizedek priesthood. It holds the right of presidency, and has power over all the offices in the Church, in all ages of the world, to administer in spiritual things, and has a right to officiate in all offices of the Church.

The second priesthood is called the Aaronic, because it was conferred upon Aaron and his seed, throughout all their generations. It is secondary to the Melchizedek, and has power to administer outward ordinances.

The bishopric is the presidency of this priesthood, and holds the keys or authority of the same.

No man has a right to this office, to hold the keys of this priesthood, *except he be a literal descendant of Aaron*. But, as a high priest of the Melchizedek priesthood, he has authority to officiate in the office of bishop, when no literal descendant of Aaron can be found—provided he is called, set apart, and ordained by the presidency of the Melchizedek priesthood.

The power and authority of the higher, or Melchizedek, is to hold the keys of all the spiritual blessings of the church—to have the privilege of receiving the mysteries of the kingdom of heaven—to have the heavens opened unto them—to commune with the general assembly and church of the first born, and to enjoy the communion and presence of God, the Father, and Jesus, the Mediator of the New Covenant.

The power of the Aaronic priesthood, is to hold the keys of the ministering angels—to administer outward ordinances—the *letter* of the Gospel—the baptism of repentance, for the remission of sins, agreeable to the covenants and commandments.

Of necessity, there are presidents growing out of, or appointed from among, those who are ordained to the several offices in these two priesthoods of the Melchizedek.

Three presiding high priests, chosen by the body, appointed, ordained and upheld, by the confidence, faith and prayer of the church, these form a quorum of the presidency. There are also twelve apostles, or travelling counsellors, especial witnesses of the name of Christ, in all the world; thus differing from other offices in the church, in the duties of their calling. They also form a quorum equal in authority, to the "three presidents."

The "seventy," are also called to preach the Gospel, and to be the especial witnesses unto the Gentiles, and in all the world, thus differing from other officers in the church, in the duties of their calling. They also form a quorum, equal in authority to the "twelve apostles," and the "three presidents."

Every decision made by either of these quorums,

must be by the *unanimous voice* of the same—that is, every member in each quorum must be agreed to its decisions, in order to make their decisions of the same power or validity, one with the other: a majority may form a quorum, when circumstances render it impossible to be otherwise. These decisions are to be made in righteousness, in holiness and lowliness of heart, meekness and long suffering, and in faith, virtue, and knowledge, temperance, patience, godliness, brotherly-kindness and charity.

In case of an unrighteous decision, it must be brought before a convention of the several quorums, which constitute the spiritual authorities of the church—otherwise there is no appeal.

The Gospel is first to be preached unto the Gentiles, secondly, to the Jews.

If a president of the high priesthood transgress, he shall be tried before twelve counsellors of that body, and their decision concerning him shall be binding. Thus none shall be exempted from the justice, and the laws of God, that all things may be done according to truth and righteousnes. The duty of the President (Brigham Young), is to preside over the whole church, and to be like Moses. Behold here is wisdom! to be a seer—a revelator—a translator—and a prophet—having all the gifts of God, which he bestows upon the head of the Church.

These form a principal part of the ecclesiastical polity of the Mormon Church of latter day saints.

The above are extracts from the "doctrines and covenants" of the Mormons.

Polygamy is practised to very great extent among

AUTHOR'S VIEWS ON POLYGAMY. 151

the high-priests and officers of the church. There are thousands of the Mormons, however, who reprobate, and disapprove of it.

The following questions seem to suggest themselves as bearing upon the polygamy practised by the Mormons. What is their rational plea from revelation as true believers? Is such a system in conformity thereto—with right reason, and with the requirements of civilized society? Will it improve the physical powers of man; impart additional mental energy, and increase the period of human existence? Is it calculated as a wise providence intended, to perpetuate his species? Does it harmonize with the *requisites* of peace and justice, and the *good order* essential to the happiness of all? In my limited reading of the Scriptures, I find nothing to sanction such a course; on the contrary, there stands at the offset of the creation a negative prohibition in Gen. ii. 22: "And the rib* which the Lord God had taken from man, made he a woman, and brought her to the man." Verse 23d of same chap.—"And Adam said, this is bone of my bone, and flesh of my flesh, she shall be called woman, because she was taken out of man." 24th: "Therefore shall a man leave his father and his mother, and shall cleave unto his wife, and they" (*the two*) "shall be one flesh." It is plain, that if more had been required for the purposes of true connubial love and happiness, and of procreation, it would have been given him, or so advised. Let us look at the 13th verse, 6th chap. of Gen.—"In the self same day entered Noah, and Shem, and Ham, and Japhet,

---

* Let them give a rib for every additional **wife.**

the sons of Noah, and Noah's wife, and the three wives of his sons with them, in the ark." It is plain, at least, in this instance likewise, that Christian bigamists have but little cause for exultation, for it is doubted whether actions of a similar character to that which the Mormons profess, was not *one* of the prominent vices that occasioned the Deluge. See 6th chap. Gen. from 1st to 7th verses, inclusive. Yet they say that they have Scripture authority! Why, King Solomon had 700 wives and 300 concubines! But was this evil habit sanctioned by any requisite as regarded his standing as a wise King of Israel? or was it done for the service of the Most High? for we read in Deut., xvii. 14 : "When thou art come unto the land which the Lord thy God giveth thee, and shall possess it, and shalt dwell therein, and shalt say I will set a ruler over us." 17. "Then shalt this ruler not multiply wives to himself, that his heart turn not away : neither shall he multiply unto himself silver and gold."

"King Solomon's wives turned his heart." "That his heart was not perfect with the Lord his God." They can have Scriptural authority (another mantle of purity for their profound consideration) for King David's adultery with Bethsheba—but, alas ! for human frailty. If we look to those nations where bigamy, or plurality of women prevails, we see men both physically and constitutionally enervated—effeminacy of character, and little or no desire to cultivate those sciences which it is designed that the human mind should grasp. They stand still, and have done so for centuries. In contemplating its natural results, and its unhappy tendencies, we are brought to consider the causes that originated, or, more properly, engendered the evil, and we are not at a loss to see that it proceeds from habitual and pampered indolence, unreasonable and carnal indulgences, unbridled passions, and the consequent inability of the intellect to discover this moral failing. What would be the consequence, if all the numerous classes of animated beings (other than man), in the particular of *regeneration*, were unrestricted by the wise ordination of *their instinct?* The answer would be—frightful havoc and total extinction of their identity. So would it be with man, if his reasoning faculties had not been vouchsafed to him. Thus is he endowed with that quickening judgment to know right from wrong; and we have demonstration that these powers of the mind are rendered lanquid, and often totally destroyed, by the brutal excesses of the sensualist; and no better term can be applied to the bigamist.

> "O! had they but the instinct of the dove,
> And could love as well."

We will suppose for an instant that this evil was prevalent throughout the earth; and, for example, let us take a community of 50,000 men and 50,000 women, and class them at a very low estimate, and let us see how it would work; we take it for granted that the number of the two sexes are equally divided:

```
 2,000 men take 10 wives each, 20,000 women, which together are 22,000
 2,000  "      5    "          10,000   "        "       "      12,000
 1,000  "      2    "           2,000   "        "       "       3,000
18,000  "      1    "          18,000   "        "       "      86,000
------                         ------                          ------
23,000                         50,000                          73,000
```

Here we have, by this moderate calculation of the

above community of 100,000, 27,000 men wifeless. The natural consequence of such a state of things will lead, in the first place, to discontent, which would grow into jealousy; to tumult, ensanguined and civil controversy, moral depravity, and disorganization of all its elements. It might be said that a very great augmentation will ensue with the growing offspring, but the fair presumption is, that male and female will be equally divided. The 27,000 men would, in any case, have to wait for their chance of getting a wife, or wives, until the young folks become marriageable, which would be at the least fifteen or twenty years (something to try their patience), and most likely then be forestalled by the more youthful swains.

And it is here, in its midst, we must look for confusion and the clashing of that near consanguinity, or relationship of blood, which is considered both a divine and moral impediment to marriage; and in such a motley community, where could be found the purity of domestic intercourse—the sanctity of true affection—the pillow of female delicacy?

In every view we take of polygamy, it is a false and vicious system, neither to be reconciled with revelation, with nature, or with reason. It is destructive to society, and to all human progress.

CHAPTER XXIV.

Grand Ball at Salt Lake City—Etiquette—Culinary Preparations—Cost of Entertainment—Author opens the Ball with one of the Wives of the Governor—Beautiful Women—Waltzing and Polkas Prohibited—Mrs. Wheelock—The "Three Graces"—Extraordinary Cotillion—Mormon Wedding—Spiritual Wives—Favorable Impression of the Public Social Life of the Mormons.

BALL AT SALT LAKE CITY.

Towards the end of April, 1854, about ten days previous to the departure of Governor Brigham Young, on his annual visit to the southern settlement of Utah, tickets of invitation to a grand ball, were issued in his name. I had the honor to receive one of them.

If the etiquette of dress, which is a necessary preliminary to the "*entré*" of her Majesty's drawing-room, had been insisted on in the vestibule of Gov. Young's ballroom, the relation of the following incidents would never have emanated from my pen.

When I arrived at the great city of the Mormons, I was clad in the tattered garments that I had worn for six months, on the journey across the Rocky Mountains. In vain I applied to every store in Salt Lake City for suitable clothes; a pair of black pants or a broadcloth coat was not to be purchased. I, however, succeeded in having a pair of stout cassimere pants made for my intended journey to California; and a gentleman by the name of Addoms, a merchant from Cedar street, N. Y.

contributed a new coat from his wardrobe. I was indebted to him also for a great deal of kindness and attention during my illness.

With my striped cassimeres, black frock coat, and a white vest borrowed for the occasion from Capt. Morris, "*en règle*"—I was as fashionably attired as any one whom I met during the evening. My friend, Egloffstien, was also invited, but there were no clothes in the city of Salt Lake to fit him; he had grown so fat and corpulent, that ready-made clothes of his size, would have been unsaleable, consequently, he declined going.

During the day, extensive culinary preparations were being made at Mr. E. T. Benson's house, where we messed. Mr. Benson had four wives; they were, on this occasion, all engaged; one making pastry and cakes, another roasting and preparing wild geese and ducks, and garnishing fat hams, etc., while the others were selecting the garments which were to be worn by the ladies on this interesting occasion.

I could not exactly perceive why such extensive cooking preparations were making; on enquiry, I learned that in this isolated city, thousands of miles from civilization, and buried, as it were, in the mountains, it was a very expensive thing to prepare a supper for a large company, at the cost of a single individual. Sugar was worth 75 cents per pound, and very scarce; sperm candles, $1.50 per pound, and everything else in proportion. It was expected, and understood, that all families who were invited, should bring their own provisions, candles, etc., and contribute for the music. The Governor furnished the ball-room only.

Strangers, of course, were exceptions to the rule.

At the appointed hour I made my appearance, chap

eroned by Gov. Young, who gave me a general introduction. A larger collection of fairer and more beautiful women I never saw in one room. All of them were dressed in white muslin; some with pink, and others with blue sashes. Flowers were the only ornaments in the hair. The utmost order and strictest decorum prevailed. Polkas and waltzing were not danced; country dances, cotillions, quadrilles, etc., were permitted.

At the invitation of Gov. Young, I opened the ball with one of his wives. The Governor, with a beautiful partner, stood *vis-a-vis.* An old fashioned cotillion was danced with much grace by the ladies, and the Governor acquitted himself very well on the " light fantastic toe."

I singled out from among the galaxy of beauty with which I was surrounded, a Mrs. Wheelock, a lady of great worth, and polished manners; she had volunteered her services as a tragedienne, at different times during my visit to Salt Lake, at the theatre, where she appeared in several difficult impersonations; I think she excels Miss Julia Dean in her histrionic talent. I had the pleasure of painting Mrs. Wheelock's portrait in the character of " Pauline," in " Claude Melnotte." She was the first wife of her husband, whom she married in England, about eight years before; her parents, who are estimable people, came over after they had embraced Mormonism. When this lady married, the spiritual wife system, had not yet been revealed.

Mr. Wheelock is a president of the seventies, and has travelled a great deal in the capacity of missionary; he had, at this time, three wives, the last one visited the ball as a bride; I was introduced by Mrs. Wheelock senior, to all of them; they looked like the three graces as they stood in the room, with their arms enfolding each other like sisters; they dwelt together in one house, and the most perfect harmony and affection seemed to exist between them. The last wife was a young girl of seventeen, well educated, and possessing great personal advantages; her parents and brothers reside in the city. I was invited to the wedding, but was prevented attending from the reason I have before assigned. I requested permission to dance with one of them; Mr. Wheelock took his new bride, and the cotillion was formed of his three wives and another lady, with their respective partners. It was a most unusual sight to see a man dancing in a cotillion with *three wives*, balancing first to one, then to the other; they all enjoyed themselves with the greatest good humor.

The particulars of the wedding, I had from a lady who was present. It seems that it is necessary before a man can take a second wife, that his first wife should give her consent; if she refuses, he is prohibited from taking another. In this case, the first wife's consent was obtained; I will not presume to say whether willingly or unwillingly; Mrs. W., the elder, possessed great good sense, and her mind was highly cultivated. It may be, she made a virtue of necessity, and yielded the assent on which her future domestic happiness depended, with a good grace.

She acted as godmother, and gave away the bride. I think on this occasion the Governor performed the ceremony. The second Mrs. Rose Wheelock is a transcendently beautiful woman. There is nothing prepossessing in the appearance of her husband, and it is a mystery to me, how he could have gained the affections of so many elegant women. Mr. W. was appointed to a mission to Great Britain previous to his last "sealing,"* and left for the States the day after the ball, he only enjoyed his last wife's society about four days—a very short honey-moon!

The lady could have married a more eligible man—She must return to her parents' house to reside, for the three years her husband would be absent; yet she preferred to be the third wife of a man she loved, and who bore a high character for morality, etc., to being the first and only wife of an inconsiderate youth.

After several rounds of dancing, a march was played by the band, and a procession formed. I conducted my first partner to the supper room, where I partook of a fine entertainment at the Governor's table. There must have been at least two hundred ladies present, and about one hundred gentlemen. I returned to my quarters at twelve o'clock, most favorably impressed with the exhibition of public society among the Mormons.

* Sealing is the ceremony of spiritual marriage.

CHAPTER XXV.

MORMON EPISODES.

" Golightly "—His Occupation and Character—Author Patronizes Him—Mrs. Golightly —She thinks Shakspeare did not understand the Passions of Men—" Oh! Frailty, thy Name is Man!"—Affecting Incident.

THE OLD LADY'S TALE.

THERE resided in Great Salt Lake City, in the year 1854, a jolly old Scotchman, who rejoiced in the cognomen of "Golightly," he was a baker by trade, a musician by nature, and a good Mormon by practice. He made first-rate bread, biscuit, and cakes, and cooked to order splendid beefsteaks and mutton chops, as my fellow traveller Egloffstien and myself can fully testify, for we patronized him daily in all the branches of gastronomy, for which he was famous.

His bakehouse was attached to his shop; a small house about a rod on one side, was his dwelling, and immediately back of the oven, in the open yard, was a covered wagon, which was used as the parlor and bedchamber of his old wife, and three daughters, aged respectively thirteen, fifteen, and seventeen, and a son of eleven years.

This old lady I frequently met in my visits to Go lightly's shop, sitting carefully wrapped up, on an old travelling chest near the fireplace; she appeared to be in very bad health, and seldom spoke, yet she often gave expression to deep drawn sighs. The three daughters assisted the father in making biscuit, cakes, etc.

Golightly was a well informed man, he had been a deist, a methodist, and was now a Mormon from conviction. I think I may say, that he firmly believed in the tenets of Mormonism, and in the many conversations I had with him, I inferred that his conduct was actuated by principle. He was an active member of a musical association, and performed well on the Kent bugle.

It was on an occasion when his professional services were required to attend the funeral procession of Brother Willard Richards, editor of the "Deseret News," that I happening in to partake of my usual lunch, I found the old lady sitting in her accustomed place, alone, and she appeared very much depressed; I asked her the cause of her sighs, etc., when she related to me the following incidents in her life. She was a native of Scotland, and had been married to her husband for a quarter of a century—had borne him twelve children, four of whom were still living. Her husband followed the trade of baker, in Edinburgh, where they lived very happily. She possessed in her own right the snug little house in which they carried on business; they owed no one, and were well to do in the world. One night her old man went to hear some strange Mormon missionaries preach; from that hour her troubles commenced, and they had steadily increased up to the present time.

Golightly becoming indoctrinated with the principles of Joseph Smith, had been baptized. In vain he tried to make his wife change her faith, presbyterian, in which she had been brought up. Finding that she would not consent, Golightly determined to emigrate to the valleys of Ephraim, the "land flowing with milk and honey." To this step also his wife refused to accede, whereupon he sold out his bakery and accumulating all the ready money he wanted for his purpose, left his family (not in want, for they had an income sufficient to live on), but without a protector, and took passage, along with many others, in a vessel from Liverpool, bound direct to New York.

After his arrival in New York, the company proceeded to St. Louis, up the Missouri to Independence, and thence, overland to Salt Lake City, where he arrived in good condition, and with the small means at his command, he built the shop and house in which I found him. He liked his new residence, and made arrangements for his family. He wrote to his wife, requesting her to sell off the property, and come over to the Valley, among the mountains, and join him, as he intended to spend the remainder of his days there.

When the old lady received this letter, she determined to brave all the dangers of a long voyage across the Atlantic, the perils of the mountains and prairies, and rejoin her beloved old man, with whom she had spent so many hours of happiness, and with whom she determined to end her life. With the assistance of kind friends, all her effects were converted into money, and she had just £200 with which to commence the journey.

Her three daughters and a young son accompanied her. Passing over her terrible sea-sickness and difficulties which attended her sea-voyage, she arrived in due time at New York, where she purchased "through" tickets for herself and children from one who styled himself an agent of the Railroad Company. After paying her money and taking seats in the cars, she found she had been cheated by the counterfeit agent, her tickets were perfectly worthless; the kind-hearted conductor, in consequence, gave her free passage to St. Louis, at which place she embarked on board the steamboat for Independence, to join a caravan of immigrants, who were also on the way to the "Valley."

At Independence she purchased two good horses and the wagon which was then at the door, together with all the necessary provisions and clothing for a five-months' journey. Her outfit cost her nearly all the money she had left; but not requiring to spend more before she got to the Valley, she made herself easy on that score. The continual state of excitement which she had been in from the time she sold out at Edinburgh, with her illness on board ship, superinduced by old age, etc., gave her the dropsy. Her daughters took it by turns to drive the team, and her kind fellow-travellers harnessed up the horses, and attended to the arduous duties of camp-travelling.

Suffering in mind and body, the caravan arrived at "Fort Laramie," where they met some teamsters who were on their return to the States. Our old lady, whose anxiety to embrace her husband increased, the nearer she approached the place he was in, was induced to inquire of one of these teamsters if he knew Mr. Golightly, in Salt Lake City? He answered, that he did, he had purchased his bread and crackers from him only a month ago. "Golightly and his wife were both well, and living very comfortably!"

"Surely, mon, you mak a mistake; 'Golightly' has na ither wife but me."

The man insisted that he had taken a spiritual wife.

"A 'spiritual wife'—I dinna ken the kind."

Our old lady had of course never heard, that polygamy was practised as a part of the religion of the Mormons. She treated the report of the teamster as a mistake, and supposed he meant that Golightly had hired a servant girl, to do the work of the house. Under this impression, she resumed her journey. But, poor woman, what was her sorrow and agony, to find on her arrival at Salt Lake that the husband of her youth, he for whom she had just submitted to such an unheard-of sacrifice of personal comfort, at her age; the father of her children, should have broken faith, and repudiated her! Heart-broken, and prostrated with disease she fell back in her wagon—in a swoon. Our old Trojan quickly applied restoratives, and endeavored to lift her into the house. "Na, na, my foot shall never cross the threshold of the house that contains anither wife; this wagon shall be my house, and my children's house; in that, during the howlings of the winter's blast, or the scorching heat of the summer, will I abide, until death takes me away." All the affection and love of Golightly, returned on again seeing his old wife, he fondled her, and prepared all the nourishment for her with his own hand, and succeeded in pacifying the old lady to submit to circumstances, which, when she found it was a part of the religion, she became more reconciled to.

But the old lady asked me, "Who do you think he married? Surely nabodie but our auld cook from Edinburgh; a dirty wench that I turned out of my house for impertinence; she followed the old man, and induced him to marry her, telling him that I never intended to come out to him. I have never set my eyes upon her, for she takes good care not to come where I am. It is now more than two years since I arrived, and the preachers have told me that if I would be baptized, I would feel perfectly contented." To please the old man, whom she still loved, she consented, and was immersed in water over her head, on a bitter cold day—but she resumed: "I canna see ony different now, I am only the worse in the body."

Her daughters are kind and affable girls, they are the sole companions of the mother, who never goes any where out of her wagon, but into the shop.

I saw Golightly several times after the revelation of his wife, he said it was an "o'er true tale," but his wife ought to know that he did not desert her, he sent for her, and loved her now more than ever, that he only took a spiritual wife, to ensure her eternal salvation; and also in accordance with his firm convictions, that he was doing right. I took the physician who was attached to the Gunnison expedition to see her, but he pronounced her case hopeless; and I would not be surprised, if ere this she is in that happy country, "where the wicked cease from troubling, and the weary are at rest."

CHAPTER XXVI.

MORMON EPISODES.

Extraordinary Abuses of the Spiritual Wife System—Fanny Littlemore—The Writer paints her Portrait—Her early Life—Attempt by her Parents to force her to marry her Uncle at Nauvoo—Her Escape to St. Louis—She writes to her Lover—Terry Littlemore—Marriage—Extraordinary Letter—Fanny's Mother exchanges Husbands with her Aunt—Her Father also exchanges Wives with her Uncle—Fanny's journey to Salt Lake—Terry Littlemore becomes a Mormon—Fanny opposed to Mormonism—Her two Sisters become spiritual Wives of a distinguished Mormon—She meets her Father and Mother in Salt Lake—The Writer becomes acquainted with her Mother and Uncle—His Journey to Parowan with them—Verification.

THE following facts were related to me by a lady residing in Salt Lake City, being interwoven with her life. I give all except the real names of the parties. This history was volunteered during the time I was occupied in painting her own, and her husband's portrait; I was not bound to secrecy, the parties immediately interested, are all residing at present in Utah. I became afterwards personally known to them, on my journey to Parowan.

Fanny Oldham, the heroine of our story, was one of several daughters. Her parents were originally presbyterians, to which faith she had been brought up. A few years previous to the commencement of this tale, her parents, as well as other members of her family, became Mormons. The scene opens in Nauvoo, in the year 1842.

" My parents resided in Nauvoo; my aunt being confined, at her own house, with a newly born infant, permission was requested of my parents, that I should go there on a visit, to assist in the domestic duties of the family, during her illness; they consented, and I, favorably impressed with my aunt's former kindness, willingly went. At this time, I was in my seventeenth year, and although surrounded by Mormons, and hearing nothing else but Mormonism preached, I still retained the religious views in which I had been educated, and refused to be baptized in their faith.

"Several days elapsed, after I was domiciled in my aunt's residence, during which time, most marked attentions were shown me by my uncle, my aunt's husband; he would affectionately kiss me for good night, and morning, and I returned his embraces with the affection of a niece. One morning, after my duties had been completed, I went into the parlor, and seated myself on the sofa; shortly afterwards my uncle came in, and taking a seat next to me, placed his arm rather familiarly around my waist, and pressed me towards him. This unusual demonstration annoyed me, and I endeavored to extricate myself from him, but he held me the tighter, and attempted to kiss me. Highly indignant at this proceeding, I asked him how he dared to treat me thus? he replied, 'that as I was to be his little wife, he thought himself privileged to kiss me.' I had never heard of the spiritual wife system, and I could not but believe he was joking; but he told me in earnest, ' that I was destined by the Almighty to become his wife." I rushed out of the room in tears, and putting on my bonnet, hurried home, as fast as I could. My father, was a man of high temper, and quick to resent an offence; he was at home when I got there, but, not daring to tell him of the insult, which had been put upon me, I went to my mother's chamber, and bursting into tears, revealed to her the scene which had just been enacted at my aunt's house."

Terry Littlemore was a cousin of Fanny, there had been a reciprocity of sentiment existing between them for years, and their troths were pledged to have been married ere this, but from the opposition they had received from Mr. and Mrs. Oldham, Fanny's parents. Terry was well to do in the world, his moral character unexceptionable, and he could not conceive the reason that he was refused the hand of his cousin Fanny. Finding that his business required his services in a town in Missouri he bade adieu to Fanny, promising to return in a few weeks and marry her even without the consent of her parents. It was during his absence that the scene I have related, took place.

Mrs. Oldham quietly listened to her daughter, and then told her that "the Prophet Joseph Smith had received a revelation from Heaven, that certain Mormon priests, were to take to themselves spiritual wives, in addition to the one wife they might have. Joseph Smith had lately seen Mrs. Oldham, and had approved of her daughter Fanny, as a wife for Mr. Wilson, Fanny's uncle, and believing as she did in the truth of Joseph Smith, she also approved of the marriage, and forbade her ever to think any more of her cousin, Terry but to prepare herself to marry her uncle in a few days.

Fanny became horror struck. She had hoped, on the bosom of a fond mother to have wept away the recollection of the unnatural and revolting proposal that had been made to her, but what was Fanny's dismay at hearing such a decision from her mother. As a last resort she sought her father, and on her knees begged him to interfere and prevent the dreadful sacrifice which was awaiting her. Her father, stern and inflexible with fanatic zeal, gave her no hope. He also approved of the marriage, and commanded her to submit, or he would use force. Poor Fanny had just time to reach her chamber when she fell fainting on the floor. When she recovered from her swoon her youngest sister was bending over her, applying restoratives. Neither of her parents had been near her for the two hours she had remained insensible.

Fanny, at the time I saw her, was the most beautiful woman in Utah. Her eyes were dark hazel, a classical nose, high forehead and luxuriant black hair. Her teeth were beautifully white, while her lips and mouth were "rich with sweetness living there." She was the mother of two children, and was 28 years of age. Still she was an elegant woman—what must she have been at the commencement of our story?

This melancholy and horrible scene had passed so rapidly before her, that she had scarcely time to realize her situation. She determined to fly. It was therefore with an aching heart that she surveyed her beautifully arranged chamber for the last time.

There were no tasseled curtains, or luxurious carpets, no hanging chandeliers, or gilded looking-glasses, but her bed was covered with linen as pure as her own spotless breast, and the primitive furniture was adorned with embroidered covers made by her own hands. A sweet little canary, the gift of dear Terry, sent forth a burst of melody when she approached his cage.

Unbidden tears streamed down her pallid cheeks, and with an unnatural composure she arranged a little bundle of clothing, which she required on her voyage. Swallowing a cup of tea which her sister had brought to her, she nerved herself for the trials she was about to encounter from the wide world She intended to claim the protection of a married sister, who had lived at St. Louis. She told her young sister, who was ignorant of what had transpired, that she intended returning to her aunt's, and kissing her affectionately, she bade her adieu.

When she got out of the house, it was near ten o'clock at night; turning towards the steamboat wharf, she flew down to the boat, and entering the cabin, she told the captain, who was well known to her, that some urgent business demanded that she should go by the first opportunity to St. Louis, and requested him not to inform her family that she was on board. The steamer left the next day, and in good time she arrived at St. Louis.

Fanny, on reaching St. Louis, immediately repaired to her sister's, who was astonished and unprepared for her arrival. She pressed Fanny to her heart, and wept from very sympathy. Poor Fanny, resting on her sister's bosom, related what had transpired at Nauvoo. Her sister determined to protect her at all hazards, and save her from the horrible fate that awaited her.

With the sanction of her sister, Fanny the next day wrote to her lover, Terry Littlemore, requesting him to come immediately to St. Louis. In the mean time, she applied herself to her needle, and earned a sufficiency to support herself.

In the course of a week, Terry Littlemore arrived at St. Louis, and hastening to his cousin Louisa's house, was soon in the arms of his beautiful betrothed. She related to him the occasion of her flight from Nauvoo, and then told him she was ready to become his wife, at any moment. Terry, fearing that his uncle would pur-

EXTRAORDINARY DEVELOPMENTS. 171

sue Fanny so St. Louis, as soon as he knew her whereabouts, determined to marry immediately, and the next morning they were united in the bonds of wedlock.

Terry Littlemore was advised to commence business in St. Louis, which he did. He opened a grocery store in partnership with another man, and furnishing a house comfortably, he took home his lovely bride. Fanny wrote to her parents, after her marriage, informing them of the fact, that they as well as her uncle might know that she was under a husband's protection.

Fanny and her husband lived happily and comfortably. In course of time she presented him with a son.

After they had been married some time, she received the following letter from her mother:

Dear Fanny:

You will be surprised to hear that after living twenty years with your dear father, and bearing him nine children, that we should be separated forever in this world. It was "revealed to both your father and myself by an angel from heaven," that we should separate, as he could not secure my eternal salvation! Your uncle, whose wife you ought to have been, has been " sealed," to me, as my Spiritual husband, and your father has been " sealed " to your aunt. I have the future care of your uncle's children, and he has the charge of your father's. Both of our families are now making arrangements to go across the plains, into some valley beyond the mountains, to seek a future and permanent home, where I hope to see you some of these days. I pray you to receive the farewell of

Your affectionate mother.

On receipt of this extraordinary epistle, Fanny hastened to her sister Louisa, who had also received a letter, conveying the same intelligence. They threw themselves into each other's arms, and wept over the infatuation and fanaticism, which had branded their parents' names with infamy.

Terry Littlemore was offered the lucrative situation of wagon-master, to conduct one hundred wagons and teams, laden with merchandize, etc., from Independence to Salt Lake. Terry decided to go, and leaving his wife and child in the care of his cousin Louisa, and his business in the joint charge of his wife and partner, took command of this expedition, and after a long journey, arrived safely in Great Salt Lake City, where another uncle held a high position in the church of the latter day saints. Here the future prospects for Terry were bright, and a fortune seemed within his grasp; he was offered by his uncle, that if he would bring his family out, that he would build him a flour mill, and give him a large tract of ground, besides stock, etc. This offer was most tempting to Terry; he determined to accept it, and making the necessary arrangements with his uncle, returned home for his family. Fanny at first declined going, but an offer having been made of a very lucrative character, to her sister's husband, which they determined to accept, Fanny not wishing to remain alone, and her husband being resolved to go, she made a virtue of necessity, and acquiesced in his wishes, although she had her fears that she was taking a wrong step.

Terry Littlemore dissolved partnership, and found he had sunk half the amount he had put in his business, by the carelessness and mismanagement of his partner.

Both families made preparations to travel, and early in the spring of 1849 they started, and with the usual adventures of a journey across the plains, arrived safely in Great Salt Lake City.

When Fanny arrived, her uncle and family called on her, and conducted her to a comfortable residence.

She was some months in the city before she would consent to see her mother, who was residing with her Uncle Lorenzo, as *husband and wife*. Her father, having had some disagreement with this spiritual wife, left her, and when Fanny arrived he was very badly off. At the time Mrs. Littlemore related to me these extraordinary episodes in her life, her father was caring horses and cattle on the pasturage beyond the River Jordan, in the Salt Lake Valley. She is now on affectionate terms with her mother. Her husband, Mr. Terry Littlemore, became a Mormon, and was baptized into the faith of the latter day saints. Mrs. Littlemore never became one. She told me her husband will never bring home a spiritual wife while she lives. Her two sisters are spiritual wives of their uncle, who is one of the great lights of the Mormon church. She seemed happy and contented, and enjoys herself. She has all the comforts, and many of the luxuries of life, and her husband is devoted to her; they live on Mill Creek, some few miles from the city, and Terry is proprietor of a flour mill, and as well-cultivated a farm, as fine teams of horses, as choice stock, and as beautiful and lovely a wife, as any man in Utah.

I subsequently learned some of the above facts from other sources. Mrs. Littlemore told me very nearly the words and substance of the foregoing, voluntarily. I think I remarked that I would write a romance, but the recital of the facts are as tragic, and as improbable as the most improbable romance that ever was written. "Truth is stranger than fiction."

CHAPTER XXVII.

Arrival of the California Mail—Murder of Mr. Lamphere by Indians on Santa Clara—Springs—Singular Phenomenon—Hot and Cold Springs—Mica—Sulphur—Plumbago-Rock Salt—Death of Willard B. Richards—Heber C. Kemball—Welsh Colony—Lieut. Beckwith's Departure for California.

April 16th.—This morning, Messrs. Atwood and Murray arrived with the California mail. They report that one of their party, a Mr. George Lamphere of Chicago, was shot by the Indians, between the Santa Clara and Rio Virgin (Virgin River). It seems that Atwood and Murray saddled their horses and prepared for their day's journey, before Mr. Lamphere had finished his breakfast. They mounted and started, intending to ride slowly along. About an hour after leaving camp, they saw Lamphere's horse galloping, riderless, towards them; as he approached, they perceived three arrows sticking in his side. They immediately suspected that their companion had been ruthlessly murdered by the Indians. They succeeded in catching the frightened horse, and secured him to a tree; afterwards they gallopped at full speed towards their late camp. They were well armed, and although they were ignorant of the force of the enemy which might be in ambush waiting for them, nothing daunted, they dashed forward, and found the dead body of their friend and companion on the road, pierced with a dozen arrows, completely stripped of all his clothing. Mr. Lamphere had a large amount of money with him, besides valuable specimens of gold, which he had obtained in Calitornia—a gold watch, etc. Everything had been stolen by the Indians of the Santa Clara.

The situation of Murray and Atwood was most critical, as evidently a large force of Indians were in the neighborhood. They recommenced their journey, and travelled at full speed until noon; encamped, and rested their animals until dark. They made a large fire, so as to show the Indians where their camp was, and, at a killing pace, journeyed all night. The Indians followed them at a distance, with a view to massacre them during the night. When they saw the smoke of the camp fire, they also encamped; and as their usual hour of surrounding a camp was just before day, when men are supposed to sleep soundest, they also rested from their fatiguing ride; but the next morning the birds had flown, and were forty miles distant from them. These gentlemen arrived at Parowan, with their animals perfectly lame, and useless for continuing their journey to Great Salt Lake City. They there procured fresh ones, and arrived safely. From their own lips, I heard the recital of the above melancholy catastrophe.

I was about to travel over this same road, and was fully alive to the dangers which might beset me; but I had to get to the sea-board, and as the party with whom I intended to travel were well armed, and composed of twenty-three able-bodied men, I felt just as secure as I would have felt on any other line of road.

PHENOMENON OF A HOT AND COLD LAKE.

About ten miles north of Salt Lake City, there are two springs close together, one salt and cold, the other fresh and hot; these springs unite at some distance, and form a lake of 400 feet in diameter—one portion of the water is hot, and the other cold, and is so all the year round.

It was said by the gentleman who described them to me, that he bathed in this lake, and that one part of his body was in the cold water, while the other was in water quite hot.

* * * * * * *

In the mountains around Salt Lake City, mica is found in large masses. I saw one block in the city, several feet square, which was perfectly transparent. It is used as a substitute for window-glass, in some of the houses of the Mormons.

Plumbago of superior quality is found on Coal Creek; and saleratus is procured in quantities from Juab Valley. Alum and sulphur abound in the different valleys of Utah.

* * * * * * *

The death of Willard B. Richards, one of the chief members of the presidency, and editor of the Deseret News, threw a gloom over the whole community. I attended his funeral. His excellency the Governor, was too unwell to officiate, but several funeral sermons were preached at the house. He was one of the earliest, and most valuable members of the church of the latter day saints.

Mr. Richards left quite a number of widows, I could

not ascertain exactly how many, but I was credibly informed by a Mormon lady, that she knew six.

Heber C. Kimball, the next in rank to Brigham Young in the church, is a noble looking man, over six feet, and well proportioned, he speaks fluently, his language is inornate, and indicates an original mind, without cultivation. He is said to have more wives than any man in Utah—the Governor not excepted.

I learned from a niece of the Governor's, that she knew personally nineteen of his wives, although he had many more.

The Governor had at the time I was in the city, thirty-three children, including several grown men and women, by his first wife, who is still living with him. I was introduced by his excellency, to eleven of his wives, at the different times I visited his residence—all of them are beautiful women. Parley Pratt introduced me to his household, I numbered five or six females, I think he has but six wives.

Ezra T. Benson, one of the apostles with whom I boarded, has four wives, three are living in the same house with him, and one in a small house, a couple of rods away. He has children by all of them, and they all seemed to live very harmoniously together. I had several conversations with these ladies on the spiritual wife system, they submit to it because they implicitly believe it to be necessary to their salvation. They argue, " Cannot a father love six children ? why can he not love six wives ?" I must say, that during a sojourn of near three months in Salt Lake City, I never observed the slightest indications of improper conduct, or lightness, amongst them—neither by conversation or otherwise. Their young ladies are modest, and unassuming, while their matrons are sedate and stately. Polygamy is by no means general, there are hundreds of Mormons who have only one wife.

WELSH SETTLEMENT IN THE MOUNTAINS.

Indian Walker, an Utah chief, relates, that on two low mountains, situated between the Red and Grand Rivers, there is a colony of white people, who live in rough stone houses, two stories high, with no windows in the lower story, and accessible only by a ladder.

These people have an abundance of sheep, and some cattle; they raise grain on the base of the mountain— this statement is corroborated by other Indian testimony.

Brigham Young says, in reference to the above, that he believes them to have been originally Welsh families, who emigrated many years ago, before the settlement of this country. He told me that he intended to send a company of Mormons to search for the colony.

May 6th.—The exploring expedition of the late Capt. Gunnison, now under the command of Lieut. Beckwith, with an escort of twenty-four mounted dragoons, under the command of Capt. Morris with orders from the government, left this morning, to explore for a pass in the Sierra Nevada Mountains, on a parallel with Great Salt Lake City.

My old *compagnon de voyage*, Egloffstien, accompanied them as engineer.

CHAPTER XXVIII.

Departure from Great Salt Lake City—Equipments for the Journey—Author Paints Portraits of Gov. Young and Apostles—His Restoration to Health—Snow Storm—Cotton Wood Settlement—Willow Creek—Lehigh—Utah Lake—Snow Storm—Pleasant Grove—Provost—Payson.

HAVING determined to go to California by the Southern route from Great Salt Lake City, through the settlements, and over the trail of Col. Fremont of 1843, which I wanted to illustrate with views, etc., I took advantage of the opportunity which offered on the 6th May, 1854.

Twenty-three Mormon missionaries, under command of Parley Pratt, were about to proceed over this route to San Bernandino, thence to San Pedro, and the Sandwich Islands; at which latter place their religious labors were to be exercised to convert those benighted islanders to the truths of Mormonism! It was the season that his excellency the Governor usually paid his annual visit to the different settlements at the South. He had also made extensive preparations for a treaty of peace with the Indians under the chieftainship of Wakara. He proclaimed his intention of accompanying Parley Pratt and his missionaries to Cedar City, the most southern settlement, a distance of 300 miles from Great Salt Lake City.

At the invitation of Gov. Young, who seemed anxious that I should have a safe escort across the desert, I completed my arrangements, and decided to proceed with this party.

I purchased a superior riding mule for which I paid, including his shoes, saddle and bridle, etc., one hundred and sixty dollars. My provisions consisted of six boxes of sardines, and one hundred pounds of crackers, made expressly for me by my eccentric friend, Golightly. Luxuries, such as butter, eggs, etc., I intended to procure at the settlements below. To the kindness of Mrs. Benson, the elder, I was indebted for four pounds of brown sugar, for which I paid one dollar per pound; and two pounds of ground coffee, at the same price,—this was a favor, for I could not have procured any at ten dollars a pound elsewhere. My wardrobe had received considerable additions at corresponding prices—four dollars for white shirts, two dollars for striped cotton shirts—about four hundred per cent. on prices for the same goods at home. I determined to provide myself with all necessaries. I had some fifteen hundred miles to travel before I reached San Francisco. I found my Pandora's box most valuable on my last journey, and everything that I might require, I put in now, unmindful of the cost. On referring to my memorandum when I arrived at San Francisco, I summed up $350 as expenses of the journey. I painted several portraits in Great Salt Lake City; among them were two of Gov. Brigham Young; one of Lieut. General Wells, General Ferguson, Attorney General Seth Blair, Apostle Woodruff, Bishop Smoot, Col. Ferrimore Little and lady, Mrs. Wheelock, and several others.

The Governor's party consisted of a large number of wagons, mounted horsemen, etc. They left on the 5th of May, 1854. I not being quite ready, having to finish a picture, was not able to leave with them. Brigham Young promised to wait for me at Provost City.

On the 6th of May I mounted my mule, (having previously sent my baggage, provisions, etc., in one of the wagons), and fully armed, and equipped with pocket compass, thermometer, drawing materials, etc., I recommenced my journey, over the route I had travelled in wagons as an invalid, three months before. I was completely restored to health—I gained the enormous increase of sixty-one pounds. When I arrived at the city I weighed one hundred and one pounds, my usual weight was one hundred and forty-five; I therefore lost forty-four pounds on the journey, and regained it, with nearly twenty pounds extra. After travelling three miles I was overtaken by a severe snow storm. I stopped at the residence of Bishop Smoot, where I remained all night, and was hospitably entertained by him. It continued snowing until ten o'clock the next morning, when I resumed my journey, and arrived at Cottonwood Settlement.

This town is eight miles from Great Salt Lake City. It contains one hundred families, who own considerable stock, etc.

Ten miles further is Willow Creek settlement, containing about seventy-five families. Ten miles further south is Lehigh, a fine town, with six hundred inhabitants, three hundred head of cattle, one hundred horses, etc. Ten miles distant, is Lake City, on the American fork in Utah Valley, containing one thousand inhabitants, five hundred head of cattle, two hundred horses, and one hundred sheep.

I have taken lodgings here, and feel rather tired with my long day's ride of thirty-five miles.

Utah Valley is the next, south of Great Salt Lake Valley, and presents a magnificent spectacle from the summit of the pass by which you enter.

Utah Lake, which you can also see from the heights, is forty-five miles long, and twelve miles broad. The lake is situated on one side, to the west of the valley. The scenery, which is enlivened by the glistening waters, although grand and sublime in stupendous mountains, flowering vales, abrupt rocky descents, etc., is without timber, except on the creeks which meander from the mountains and entirely surround the valley. Sparse growths of young cottonwood are the only trees I have seen, except in the canons of the mountains, on which grow pines, cedars, and a species of mahogany.

May 8th.—I awoke this morning and found another snow storm raging, and very disagreeably cold; but if I allow these trifles to detain me, I shall not be in time to meet the Governor.

After breakfast I mounted my mule, and in an hour I arrived at Pleasant Grove, containing 300 inhabitants. Passing through, without stopping, I continued my journey, the snow blowing in my face the whole way, until I rode into Provost, a distance of ten miles from Pleasant Grove. I was disappointed in finding that his Excellency had departed that morning for "Petetnit," nineteen miles further. I stopped there to dine, gave my mule a good feed, and after warming my almost frozen feet, I jumped into my saddle, determined to ride the nineteen miles before dark. Onward I went, putting my mule to his mettle. He, not minding a gallop, tried to create a circulation. In a couple of hours it cleared up, and at six o'clock I rode into Petetnit.

Provost City is a large settlement, containing about

eight hundred and sixty families, equal to five thousand inhabitants, two thousand head of cattle, three thousand sheep, five hundred horses, several woollen manufactories and carding machines, shingle machines, two sawmills, a seminary and several schools, pottery, tannery, etc. Here are five hundred men capable of bearing arms.

Provost City is built on Provost River, which abounds in salmon trout of delicious flavor and large size.

Evan M. Green is mayor; Elias Blackburn, bishop. There are four bishops to this city.

CHAPTER XXIX.

Join Governor Young and Parley Pratt—Hospitality of the Mormons—Apostle Benson—Petetnit—Nephi—Wakara (Indian Chief)—Wakara's Camp Ground—Brigham Young's Wife—Long Caravan—Arrival at Wakara's Camp—His Refusal to meet the Governor—Treaty of Peace not Concluded—Presents of Cattle, etc., to Wakara—Grand Council of Indians and Mormons—Speech of an Old Chief—Address of a "San Pete Chief"—Wakara Refuses to Speak—He Dissolves the Council—Reassembling of the Council—Brigham Young's Address—Speech of "Wakara"—Peace Proclaimed—Calumet Smoked—Indian Capture of Children—Brigham Young's Residence.

GOVERNOR YOUNG and party were encamped at the edge of thet own of Petetnit; when I rode up, I saw the commanding person of the Governor, towering above the crowd of men by whom he was encircled. As soon as he saw me, he approached; I alighted to greet him he received me as he always did, in a most cordia manner. After selecting a person to take my mule, he gave me in charge to Mr. Ezra Parrish, with a request to take the best care of me until we were ready to start in the morning. I supped, and then went to the meeting, where I heard an eloquent and feeling exhortation to the people, to practise virtue, and morality. Apostle Benson also preached a sermon on the restoration of Israel to Jerusalem, which would have done honor to a speaker of the Hebrew persuasion; they call themselves "Ancient Israelites of the order of the Melchizedek priesthood."

These Mormons are certainly the most earnest

religionists I have ever been among. It seems to be a constant self-sacrifice with them, which makes me believe the masses of the people honest and sincere.

* * * * * *

9th.—This morning I was invited to breakfast with Governor Young and lady. On leaving the hospitable house where I had slept, the host refused to take payment for my supper and lodgings, or for the care of my mule.

I made arrangements with Parley Pratt's company, to take my provisions and bag of clothes to San Bernandino in a wagon, for "thirty dollars." The day was fine, and we started with an accession of five wagons, and several horsemen to the party. The town of Payson, or Petetnit, contains one thousand inhabitants, five thousand head of cattle, one hundred and fifty horses, five hundred and fifty sheep, two saw-mills, flour-mill, etc. It is organized as a city, enclosed with a high wall; the houses are generally built of logs and "adobes," one story high. We left Payson at nine o'clock, on the 10th May, and camped at noon, on a creek twelve miles S. S. W. from town.

The country around looks beautifully verdant, brilliant colored flowers cover the plain, and the grass is excellent. At five o'clock P. M. we camped before Nephi, which is a large town, containing six hundred men, women, and children; one hundred and fifty men bearing arms, six hundred head of cattle, and six hundred sheep, flour-mills, saw-mills, etc. Jos. L. Heywood, president, Josiah Miller mayor.

The Governor and party, were met by the authorities of the city, I was introduced to the old Patriarch Wm. Cazier, who invited me to the hospitalities of his house. Nephi is twenty-six miles from Payson. I attended meeting this morning, and Governor Young addressed the people, exhorting them to be kind and friendly to the Indians, etc. To-morrow we are to have an interview with Walker, the Utah Chief. A portion of the cattle intended for him was obtained at this place. The massacre of Captain Gunnison, by the Parvain Indians, caused great excitement among the inhabitants of the villages. The various tribes of Indians, who had, at different times, been wantonly and cruelly shot down, like so many wild beasts, by the American emigrants to California, were now incited to revenge. The first principle inculcated among them was life for life; it made no difference whether, in their wrath they massacred an innocent, or an unoffending man; "a white man slew my brother, my duty is to avenge his death, by killing a white man." Their first open demonstration, was the massacre of Gunnison; and the allied troops of Utahs, Pahutes, Parvains, and Payedes determined to continue in open hostility, both to the Mormons, and Americans. The inhabitants of the different settlements withdrew within the walls of their towns, and vigilant watchers, well armed, patrolled them all night. Major Biddell, the sub Indian agent, was sent to parley with the chief of the tribes, and succeeded in obtaining a truce, until the Governor could personally make arrangements for a treaty of peace. Preliminaries being settled, the chiefs of the tribes were to meet Governor Brigham Young, at the camp of the Wakara. We left Nephi, and arrived at noon, on the road opposite to Wakara's camp, twelve miles from town.

TREATY OF PEACE WITH THE UTAHS.

The camp-ground or village where Wakara permanently resides, when not travelling, is situated about one mile off the main road, from the city of Nephi, to the Sevier River. Gov. Young made extensive preparations for this treaty. A large cavalcade accompanied him from Great Salt Lake City, composed of Heber, C. Kimball, Woodruff, John Taylor, Ezra T. Benson, Lorenzo Young, Erasmus Snow, Parley Pratt, (his apostles and advisers), together with about fifty mounted men, and one hundred wagons and teams filled with gentlemen, with their wives and families. This was an imposing travelling party, all following in regular succession; taking the word of command from the leading wagon, in which rode Gov. Brigham Young. One of his wives, an accomplished and beautiful lady, who made her husband's coffee, and cooked his meals for him at every camp, thus making herself a most useful appendage to the camp equipage, as well as an affectionate and loving companion to her spiritual lord while travelling. I sometimes formed a third party on the road, and frequently had my seat at their primitive table, which was, in fine weather, a clean white cloth, spread over the grass; or, in rainy weather, a movable table was arranged in the wagon. Venison, beef, coffee, eggs, pies, etc., were served at every meal.

I have often stopped at the top of some commanding eminence, to see this immense cavalcade, lengthened out over a mile, winding leisurely along the side of a mountain, or trotting blithely in the hollow of some of the beautiful valleys through which we passed, to the sound of musical choruses from the whole party, sometimes ending with

> "I never knew what joy was
> Till I became a Mormon,"

to the tune of " bonny breastknots." Certainly, a more joyous, happy, free-from-care, and good-hearted people, I never sojourned among. When the cavalcade arrived on the road, opposite to Walker's camp, Gov. Young sent a deputation to inform Wakara that he had arrived, and would be ready to give him an audience at a certain hour, that day.

Wakara sent word back to say, "If Gov. Young wanted to see him, he must come to him at his camp, as he did not intend to leave it to see any body."

When this message was delivered to Gov. Young, he gave orders for the whole cavalcade to proceed to Wakara's camp—"If the mountain will not come to Mahomet, Mahomet must go to the mountain."

The Governor was under the impression that Walker had changed his mind, and intended to continue the war, and for that reason declined to meet him. But old Wakara was a king, and a great chief. He stood upon the dignity of his position, and feeling himself the representative of an aggrieved and much injured people, acted as though a cessation of hostilities by the Indians was to be solicited on the part of the whites, and he felt great indifference about the result.

Gov. Young, at the expense of the people of Utah, brought with him sixteen head of cattle, blankets and clothing, trinkets, arms and ammunition. I expressed much astonishment, that arms and ammunition should be furnished the Indians. His excellency told me that from their contiguity to the immigrant road, they possessed themselves of arms in exchange and trade, from American travellers. And as it was the object of the Mormons to protect, as much as possible, their people from the aggressions of the Indians, and also from the continual descent upon their towns—begging for food, and stealing when it was not given, he thought it more advisable to furnish them with the means of shooting their own game. The Utah Indians possess rifles of the first quality. All the chiefs are provided with them, and many of the Indians are most expert in their use.

When we approached Wakara Camp, we found a number of chiefs, mounted as a guard of honor around his own lodge, which was in the centre of the camp, among whom were Wakara and about fifteen old chiefs, including Ammon, Squash-Head, Grosepine, Petetnit, Kanoshe, (the chief of the Parvains), a San Pete chief, and other celebrated Indians. The Governor and council were invited into Wakara's lodge, and at the request of his excellency, I accompanied them. Wakara sat on his buffalo-robe, wrapped in his blanket, with the old chiefs around him; he did not rise, but held out his hand to Gov. Young, and made room for him by his side.

After the ceremony of shaking hands all round was concluded, our interpreter, Mr. Huntington, made known the object of the Governor's visit, and hoped that the calumet of peace would be smoked, and no more cause be given on either side, for a continuation of ill-feeling, etc.

For five minutes intense silence prevailed, when an old grey headed Utah chief got up, and in the effort, his blanket slipped from his body, displaying innumerable marks of wounds and scars. Stretching aloft his almost fleshless arm, he spoke as follows:—

"I am for war, I never will lay down my rifle, and tomahawk, Americats have no truth—Americats kill Indian plenty—Americats see Indian woman, he shoot her like deer—Americats no meet Indian to fight, he have no mercy—one year gone, Mormon say, they no kill more Indian—Mormon no tell truth, plenty Utahs gone to Great Spirit, Mormon kill them—no friend to Americats more."

The chief of the San Pete Indians arose, and the tears rolled down his furrowed cheeks as he gave utterance to his grievances:

"My son," he said, "was a brave chief, he was so good to his old father and mother—one day Wa-yo-sha was hunting rabbits as food for his old parents—the rifle of the white man killed him. When the night came, and he was still absent, his old mother went to look for her son; she walked a long way through the thick bushes; at the dawn of day, the mother and the son were both away, and the infirm and aged warrior was lonely: he followed the trail of his wife in the bush, and there he found the mother of his child, lying over the body of Wa-yo-sha, both dead from the same bullet. The old woman met her son, and while they were returning home, a bullet from the rifle of Americats shot them both down." He added, "old San Pete no can fight more, his hand trembles, his eyes are dim, the murderer of his wife, and brave Wa-yo-sha, is still living. San Pete no make peace with Americats."

The old warrior sank down exhausted on his blanket.

Wakara remained perfectly silent.

Gov. Young asked him to talk, he shook his head,

"No," after the rest had spoken, some of whom were for peace, Wakara said, "I got no heart to speak—no can talk to-day—to-night Wakara talk with great spirit, to-morrow Wakara talk with Governor."

Gov. Young then handed him a pipe, Wakara took it and gave one or two whiffs, and told the Governor to smoke, which he did, and passed it around to all the party; this ended the first interview.

An ox was slaughtered by the orders of Gov. Young, and the whole camp were regaled with fresh beef that evening. I made a sketch of Wakara during the time that he sat in council. I also made a likeness of Kanoshe, the chief of the Parvain Indians.

The next morning the council again assembled, and the Governor commenced by telling the chiefs, that he wanted to be friends with all the Indians; he loved them like a father, and would always give them plenty of clothes, and good food, provided they did not fight, and slay any more white men. He brought as presents to them, sixteen head of oxen, besides a large lot of clothing and considerable ammunition. The oxen were all driven into Wakara's camp, and the sight of them made the chiefs feel more friendly.

Wakara, who is a man of imposing appearance, was, on this occasion, attired with only a deer-skin hunting shirt, although it was very cold; his blue blanket lay at his side; he looked care-worn and haggard, and spoke as follows:

"Wakara has heard all the talk of the good Mormon chief. Wakara no like to go to war with him. Sometimes Wakara take his young men, and go far away, to sell horses. When he is absent, then Amerecats come and kill his wife and children. Why not come and

PEACE WITH THE INDIANS CONCLUDED. 193

fight when Wakara is at home? Wakara is accused of killing Capt. Gunnison. Wakara did not; Wakara was three hundred miles away when the Merecat chief was slain. Merecats soldier hunt Wakara, to kill him, but no find him. Wakara hear it; Wakara come home. Why not Merecats take Wakara? he is not armed. Wakara heart very sore. Merecats kill Parvain Indian chief, and Parvain woman. Parvain young men watch for Merecats, and kill them, because Great Spirit say— 'Merecats kill Indian;' 'Indian kill Merecats.' Wakara no want to fight more. Wakara talk with Great Spirit; Great Spirit say—'Make peace.' Wakara love Mormon chief; he is good man. When Mormon first come to live on Wakara's land, Wakara give him welcome. He give Wakara plenty bread, and clothes to cover his wife and children. Wakara no want to fight Mormon; Mormon chief very good man; he bring plenty oxen to Wakara. Wakara talk last night to Payede, to Kahutah, San Pete, Parvain—all Indian say, 'No fight Mormon or Merecats more.' If Indian kill white man again, Wakara make Indian howl."

The calumet of peace was again handed around, and all the party took a smoke. The council was then dissolved.

Gov. Young intended to visit all the settlements south, to Harmony City. Wakara told his excellency, that "he and his chiefs would accompany him all the way and back, as a body-guard." Grosepine, Ammon, Squash-head, Wakara and his wife, Canoshe and his wife, and about thirty Indian young men, all mounted on splendid horses, got ready to accompany the Governor's party. During the day, a great many presents were distributed among the tribe.

When I returned to our camp, I saw a crowd around the Governor's wagon. I approached, and found that his excellency had just concluded a purchase from the Utahs of two children, about two to three years of age. They were prisoners, and infants of the Snake Indians, with whom the Utahs were at war. When the Governor first saw these deplorable objects, they were on the open snow, digging with their little fingers for grass-nuts, or any roots to afford sustenance. They were almost living skeletons. They are usually treated in this way—that is, literally starved to death by their captors. Gov. Young intended to send them to Salt Lake City, and have them cared for and educated like his own children. I never saw a more piteous sight than those two *naked infants*, in bitter cold weather, on the open snow, reduced by starvation to the verge of the grave—no, not the grave; for if they had died, they would have been thrown on the common for the wolves to devour!

CHAPTER XXX.

Portrait of Wakara—Indian chiefs, to accompany the Expedition to Harmony City—Seveir River—Swollen Waters—Wagons ferried over—Col. Fremont—Fillmore City—Massacre of Capt. Gunnison—Parowan Indians—Kanosh—Capt. Morris—His conduct justified—Author trades for a Horse—Extraordinary Phenomenon of Insects.

We remained in camp, near Wakara's village until next day; I induced Wakara, to sit for his portrait; also Squash-head, Baptiste, Grosepine, Petetnit, and Kanoshe the chief of the Parvain Indians.

12th. We all started this morning, for the Seveir river; we arrived at the crossing at 4 o'clock P. M. and found the stream very high, and unfavorable. There had been a bridge built, a year before, but the swollen and rapid stream, carried it away; on the bank of the river, were piled up several of the planks saved from the wreck. All hands went to work to construct a raft, which they completed in an hour, and by 8 o'clock P. M., 41 wagons (the rest remained behind,) were ferried over in safety; we camped on the other side of the river.

By invitation, supped with Brigham Young: I conversed through an interpreter with Wakara, the Utah chief. He states that he supplied Jose, the Mexican, whom Col. Fremont found in the mountains, and who left at Parowan, with a mule, to go with several Indians, back on Col. Fremont's trail, to find the "cache," (the goods buried in the snow,) about 100 miles from

Parowan; he had been absent 30 days, yet nothing had been heard from them. He also told me of his interview with Col. Fremont, some years before, and showed me the place where Col. Fremont crossed the Seveir River, which was a short distance from where we crossed it. He remembered Col. Fremont, as the "great Americats Chief." While the men were constructing their raft, I occupied myself in making drawings of the surrounding country.

13*th*. We left the Seveir for Fillmore City, (called after the President of the U. S.,) which is 35 miles south of us. After travelling ten miles, we camped "to noon," giving an opportunity for the animals, to enjoy the luxuriant grass, which grows abundantly in this valley. ("Round Valley.") We arrived at Fillmore City, in Parvain Valley, Millard county, at 5 o'clock. This valley is sixty miles long and fifty miles wide; the Seveir Lake is forty miles from Fillmore. Within ten miles of the city, to the west, four fresh water lakes are to be found. Fillmore City, contains one hundred and fifty families, one thousand head of cattle, three hundred sheep, saw-mills, and flour-mills, etc., etc. A wall of adobes is built all round the city, protecting the inhabitants from the Indian aggressions.

Capt. Gunnison's party were encamped at Cedar Spring, in this valley, at the time of their massacre.

This afternoon, accompanied by two interpreters and several other gentlemen, we proceeded to the Parvain Indian's camp, to see their celebrated chieftain, Kanoshe, whose portrait I was anxious to obtain. I found him well armed with a rifle and pistols, and mounted on a noble horse. He has a Roman nose, with a fine intelligent cast of countenance, and his thick black hair is brushed off his forehead, contrary to the usual custom of his tribe. He immediately consented to my request that he would sit for his portrait; and on the spot, after an hour's labor, I produced a strong likeness of him, which he was very curious to see. I opened my portfolio and displayed the portraits of a number of chiefs, among which he selected Wa-ka-ra, the celebrated terror of travellers, anglicised Walker, (since dead). He took hold of it and wanted to retain it. It was, he said, " wieno,"—a contraction of the Spanish "bueno"— very good. I also learned from him, through the interpreters, the following facts, relating to Gunnison's massacre.

"There were about thirty Parvain Indians, encamped six miles, N. W. of Gunnison's camp, on Cedar Spring. Potter, a Mormon guide, and one of the exploring party went out to shoot ducks; one of the Parvains was also shooting rabbits, and hearing the explosion of fire-arms, he marked the direction, and followed the men to their camp. This Indian was the son of a Parvain Chief, who was killed by a party of emigrants, under command of Capt. Hildreth, about two weeks before. Marking the spot, he repaired to his own camp, and commenced to make inflammatory speeches to his tribe; he made a fictitious scalp out of horse hair, attached it to a pole, and elevating it, commenced the war dance; the rest of the Parvains continued dancing until midnight.

They were incited to revenge, for the unprovoked murder of their old chief; who, together with some women and young men, went into Hildreth's camp merely to beg food. They were ordered out, and force was used to take away their bows and arrows; in the scuffle, one of the Americans got his hand cut with an arrow-head, when they were fired upon with rifles, and several persons killed; among them this old chief.

The Parvains, before day, started for Gunnison's camp, surrounded the party who were breakfasting under cover of the willows which grew on the banks of the creek. Capt. Gunnison was the first man who had finished his breakfast; he arose, and while speaking to his men, the Indians with a tremendous yell, fired upon them. Capt. Gunnison raised his hands and beckoned them to stop. The men immediately fled, only one man fell by the first fire on the spot. The men's first endeavors were to reach their horses; the Indians pursued them, and shot them from their horses. The American party never fired a gun, the last man fell three miles from camp.

Kanoshe, the chief, was sixteen miles away from the scene of the massacre, and knew nothing about it. One of the tribe brought a horse into camp, and told Kanoshe what had transpired. Kanoshe took the horse to the Mormon settlement, (Fillmore), and gave it up to the authorities. He then proceeded to the Indian camp for the purpose of procuring the property of the slain, to render it up to the Americans. The Parvains were exasperated at his interference, and several arrows were aimed at him to kill him.

His indomitable courage alone saved him. He finally persuaded them to give up the papers and effects of the slain, which he delivered to the proper authorities. The Mormon guide was also slain.

The remains of the bodies of those who were murdered, were afterwards interred by the Mormons.

When the alarm was given to the main body of Capt. Gunnison's party by one of the men who escaped from the Indians, Capt. Morris and a detachment of his dragoons, instantly galloped to the scene of action, thirty miles off; they were totally unprepared for anything but offensive warfare.

They arrived on the spot, and found the mutilated remains of their comrades, but no signs of Indians. The weather was very cold, and the ground frozen hard; they had nothing with them but their swords, to dig into the frozen earth, and were thus compelled to leave them, until they could send from camp, men with pickaxes, etc.; besides, they were among treacherous and hidden enemies. The living men at the main camp, claimed the first duty of Capt. Morris, and as he could do no good to the dead by remaining, he retraced his steps to the main camp, to protect it from a like aggression, if attempted. He did not know but that the whole of the Indians were in warlike array around him, secretly hid away among the willows on the creek.

Some blame seems to have attached to Capt. Morris; I read an article at Salt Lake City, in a late American paper, in which his conduct was censured. I showed him this paper, and he personally explained the situation he was placed in, and told me that his duty as an officer, was to protect the lives of his surviving party, at the expense of the fraternal feelings and sympathies which he entertained for the lamented dead. I have no hesitation in saying that, from my knowledge of the circumstances of the case, Capt. Morris was perfectly justified in acting as he did.

* * * *

At Fillmore I renewed my acquaintance with Mrs. Webb, who kindly entertained me when I passed through this place three months ago.

14th. To-day I made a trade with Wakara, for a

horse; I gave him my double-barrel gun and a blanket in exchange, I have now a relief for my mule—we have a long journey before us, and I must give him as much liberty as possible. My sole dependence is on him, for crossing those dreaded jornadas* of over two hundred miles in extent.

I made several views and sketches to-day. Fillmore is 33 miles S. S. E. from the Sevier River, latitude 38° 59′

The Parvain Indians are a dirty degraded set of beings, scarcely deserving the name of human. They are much inferior to the Utahs, both in mind and appearance.

The Utahs have a large number of horses, and when mounted for a journey they are caparisoned with bells and gaudy trappings. The men paint their faces with vermilion, except when they go to war—they then paint them black. They are curiously attired in buckskin shirts, leggings, and moccasins, beautifully marked with beads and porcupine quills. They generally travel bare-headed, with sometimes a single feather in their hair. They are very fond of red and blue blankets, and use them in the manner of a Roman Toga.

PHENOMENON OF INSECTS RESEMBLING GUNPOWDER.

Riding leisurely along, at the extreme end of the caravan, I noticed on the ground, what I supposed to be gunpowder. I knew that Gov. Young had a considerable quantity with him to give the Indians, and every man had more or less, a pound—I attributed it to the acci-

---

* A journey: the absence of water and grass, makes it necessary to continue across the desert without stopping.

dental breaking of a keg, as the wagon jolted along, it might have lost through the crevices. I also noticed that the powder was only in the ruts made by the wheels of the wagons. The quantities seemed to increase, and determining to prevent, if possible, any further waste, I galloped to the other end of the train, and called Gov. Young's attention to it. The caravan was stopped, and I dismounted to obtain a specimen of it to show the Governor, when I discovered that they were minute living insects of the beetle tribe, but no larger than a grain of rifle gunpowder, and at the distance of a foot it was impossible to tell the difference. When the heaps were closely examined, they appeared a moving living mass; on the road, ahead of the wagon there were none to be seen; the weight of the wheels seemed to have pressed them through the snow, with which the whole valley was covered. The contrast of these minute, black insects on the dazzling snow was remarkable; for ten miles, it appeared as if two continuous trains of gunpowder, from three to five inches wide, were laid the whole length of the Parvain Valley. Neither the Governor nor the gentlemen who accompanied the expedition, had ever remarked a similar phenomenon before, although they had frequently travelled over the same road.

CHAPTER XXXI.

Corn Creek—Meadow Creek—Exploration of Vinegar Lake—Mephitic Gas—Sulphuric Acid—Sulphur—Alum—Volcanic Appearance of the Country—Beaver River Valley—Lieut. Beale's Pass into the Valley of the Parowan—Col. Fremont's Pass in the same Valley—Author crosses his own Trail made three Months before—His Feelings on the Occasion—Red Creek Cañon—Hieroglyphics—Granite Rocks—Remains of a Town—Arrival at Parowan—Brigham Young—Old Acquaintances.

MAY 15*th*. On rising this morning I found a snow storm raging on the mountains; in the valley it was raining, and the temperature 38°, cold enough to make great coats desirable. We left camp at 8 o'clock, and after travelling ten miles, crossed a fine stream of water called Meadow Creek, banked with willows; two miles further we crossed another rivulet, also fringed with willows and a few cottonwood trees.

The soil in Parvain Valley is rich and highly productive; the earth is covered with parterres of beautiful wild flowers, which are quite refreshing to the eye, contrasted with the snowy mountains all round us.

At 6 o'clock we camped on Corn Creek, 33 miles from Fillmore City; this is the only water from Meadow Creek, a distance of twenty-one miles.

The whole country in this neighborhood is of volcanic origin. Black cinders abound on the mountains, and a kind of grey pumice stone is found in the valleys. Sulphur in large quantities lies on the open ground in the ravines.

Mountains of pure solid transparent rocksalt rear their majestic heads in Juab Valley, a few miles south.

16*th*. Wakara, the Utah chief, one of the Indians who accompanied us, informed me that a few miles from our present camp there was a most extraordinary vinegar lake, where all bad spirits dwell; a place where a living animal never was seen, and near which there was no vegetation. Our interpreter told me he had heard before of such a lake, but he placed no faith in it. Wakara said he would go along and show us the place. Being anxious and determined to explore, and make some discovery which might benefit science, if any was to be made on this journey, I induced several Mormons to make up a party sufficiently large to insure us against an Indian surprise. The next morning we left the main trail, and proceeded about two miles in an easterly direction towards the base of the Warsatch range. Our path was covered with large quantities of obsidian, and presented every indication that the lake we were approaching was of volcanic origin. Before the lake was in sight, the atmosphere gradually became unpleasant to inhale, leaving a sulphurous taste on your palate. The approach to the lake was, for the last five hundred yards, over limestone rock, carbonized evidently from great heat, at some remote period. The air was greatly charged with sulphuric hydrogen gas, which caused me to feel an inclination to vomit. It affected the rest of the party in a similar manner. Being determined to examine further, we descended the lime formation for about one hundred feet; this brought us immediately on the spot. Its appearance indicated from the character of the surrounding country, that it evidently had been a lake; it now looked like the dry bed of what was once a lake. The surface was covered with an efflorescence to the depth of a foot, more solid, however, as you dig into it, composed of impure alum, and most probably formed by the action of sulphuric acid on feldspathic rock. Further towards the base of the mountain which bounded it on the east side, I found large quantities of pure crystalized alum, and also pure sulphur. This efflorescence which covers the lake, might be composed by the spontaneous evaporation of a mixture of sulphate of iron, and tersulphate of alumina, excess of sulphuric acid being present.

We with great caution commenced to walk over this surface, and discovered that it undulated with the weight of our bodies. I felt as if walking on thin ice, which bent, without breaking beneath my weight. As we approached the centre, we heard a roaring, which our Indian said was caused from "big fire below." I put my ear close to the earth, and was almost sure it proceeded from the escape of either gas or the passage of water. With a pickaxe, brought for exploring purposes, an orifice about a foot in diameter was dug. The axe was suddenly driven through, when a yellow, muddy liquid gushed forth in a continued stream. I tasted the liquid, when to my surprise, it was a strong acid, which immediately set my teeth on edge. Sulphuric acid in large proportions was present; this crust of over a mile in diameter, was resting on the surface of this immense body of diluted sulphuric acid. Oxide of iron in large quantities is to be found cropping out of the base of the mountains; sulphur in large quantities is also present. These materials, acted upon by volcanic heat, will produce a white powder, which partakes of the character of the substance, forming the covering to the lake. In the neighborhood of some volcanoes, sulphuric acid is found impregnated with lime and baryta, both of which are abundant on the margins of this wonderful lake. The roaring is evidently produced by the force of the liquid through some subterranean cavern; over this vast field of efflorescent sulphate of oxide of iron, there are no signs of vegetation.

On the mountains, and towards its southern boundary, some few Norway pines and cedars grow. The sulphuretted hydrogen gas which impregnates the atmosphere, prevents birds or animals from inhabiting or resorting near its neighborhood. This gas I judge to be generated by the action of diluted sulphuric acid, on proto-sulphate of iron, all which ingredients are to be found here. Feeling ill effects from inspiring this gas, I finished my examinations quickly, and sought a purer atmosphere. I made a drawing of the lake, and surrounding mountains. This extraordinary place had probably never before been examined by a white man. None of the many Mormons who were present, and to whom I related the particulars, ever explored it. It lies directly at the base of the Warsatch Mountains, in about 38° 26' latitude, and the same longitude as Fillmore City, and nearly 35 miles south of it. We rejoined our caravan at their noon camp.

About one o'clock we resumed our ride, and after a gentle ascent through a beautiful pass in the mountains, we emerged into a large and fertile valley called "Beaver Valley." We camped on Beaver River, thirty miles from Corn Creek. This stream is twenty-five feet wide, and two feet deep at the crossing; it rises and sinks alternately to the Sevier Lake, into which it empties. Only small willows grow on its banks. Beaver River abounds in wild ducks, snipe, and other water-fowl.

17th.—This morning, at daylight, there was a severe frost—water froze in camp half an inch thick. We left camp at half past seven, and after a drive of six hours, the caravan camped on Little Creek cañon—the pass through which Lieut. Beale entered Little Salt Lake Valley, a few months previously.

We harnessed up again, and in an hour crossed the trail which Col. Fremont and our party made on entering this valley from the Warsatch mountains, on the 6th of February preceding.

Under what different circumstances I travelled the same road at that time! When I turned to survey the snowy mountains among which we had suffered so much, and from the dangers of which we had been so miraculously preserved, tears involuntarily flowed from my eyes—I was completely overcome.

I made a drawing of this pass, and also of Lieut. Beale's.

On Red Creek cañon, six miles north of Parowan there are very massive, abrupt granite rocks, which rise perpendicularly out of the valley to the height of many hundred feet. On the surface of many of them, apparently engraved with some steel instrument, to the depth of an inch, are numerous hieroglyphics, representing the human hand and foot, horses, dogs, rabbits, birds, and also a sort of zodiac. These engravings present the same time-worn appearance as the rest of the rocks; the most elaborately engraved figures were thirty feet from the ground. I had to clamber up the rocks to make a drawing of them. These engravings evidently display prolonged and continued labor, and I judge them to have been executed by a different class of persons than the Indians, who now inhabit these valleys and mountains—ages seem to have passed since they were done.

When we take into consideration the compact nature of the blue granite and the depth of the engravings, years must have been spent in their execution. For what purpose were they made? and by whom, and at what period of time? It seems physically impossible that those I have mentioned as being thirty feet from the valley, could have been worked in the present position of the rocks. Some great convulsion of nature may have thrown them up as they now are. Some of the figures are as large as life, many of them about one-fourth size.

On Red Creek cañon, a mile further down the valley, there are the remains of a town, built of adobes; ancient articles of housekeeping have been found there. These remains were remarked by the first "Mormons" who came in the valley. Indians never live in adobe houses; their lodges are always of umbrageous foliage, or skins of animals.

As soon as our party were descried from the observatory at Parowan, the authorities of the town, and numbers of other gentlemen, came out to welcome the arrival of his excellency, Governor Young; and I never could have imagined the deep idolatry with which he is almost worshipped. There is no aristocracy or presuming upon position about the Governor; he is emphatically one of the people; the boys call him Brother Brigham, and the elders also call him Brother Brigham. They place implicit confidence in him, and if he were to say he wanted a mountain cut through, instantly every man capable of bearing a pick-axe would commence the work, without asking any questions, or entertaining expectation of payment for services.

He must certainly possess some extraordinary qualities, which could inspire such unlimited confidence in two hundred thousand Mormons.

We entered Parowan about five o'clock. I was affectionately greeted by those persons who administered to my sufferings some few weeks before. I had changed so much, and grown so fat, that not one of them knew me.

Mrs. Heap, my old landlady, could not believe I was the ugly, emaciated person whose face she washed only three months before.

CHAPTER XXXII.

Description of Parowan—Cedar City—Fish Lake—Iron Ore—Bituminous Coal—Future Destiny of Cedar City—Henry Lunt—Affecting Incident—Portrait of a dead Child—A Mother's Gratitude—Harmony City—Parley Pratt—Piede Indians—Personal Privations of Mormons—Bid Adieu to Gov. Young—Letter of Introduction to President of San Bernandino.

Parowan is situated immediately under a very high range of irregular, rugged mountains, fringed with timber. A fine stream of water runs through the city, which is sixty rods square, surrounded with a wall, six feet at the base, and tapering upwards to two and a half feet, the wall is twelve feet high, and extends back from the town six miles.

"The valley of the Parowan, or Little Salt Lake Valley, is about sixty miles east of the meadows of Santa Clara, between 37° and 38° of north latitude, and between 113° and 114° west longitude; elevation above the sea, five thousand feet." (Fremont's letter.)

It contains one hundred families, five hundred head of cattle, one hundred and fifty horses and mules, and three hundred sheep.

Provisions of all kinds, are very scarce and high; their supplies are procured either from Salt Lake City, three hundred miles north, or San Bernandino, five hundred miles over the deserts to the south. C. V. L. Smith is president; Lewis, bishop; John Steele, mayor.

18*th.*—The whole party left this morning at ten o'clock, for Cedar City, Coal Creek; we arrived there at two o'clock—eighteen miles to the south of Parowan.

Mr. Henry Lunt, a well informed, and generous hearted Englishman, was, it is supposed, the first white man who ever entered this valley, or the river of the Great Basin. With twenty-two men he arrived at the present site of the city, two years and a half ago to form a settlement.

Cedar City now contains one thousand inhabitants, who possess fifteen hundred head of cattle, besides a large number of horses, mules, and sheep. The city is half a mile square, and completely surrounded by an adobe wall twelve feet high, six feet at the base to two and a half at the top; the building of the wall was attended by a great deal of labor; the persevering industry of these people is unsurpassed. A temple block is in the centre of the city, covering twenty acres of ground, the building lots are each twenty rods by four rods.

Twenty miles to the eastward of Parowan, there is a fresh water lake, formed by a stream from the Warsatch Mountains, which is filled with salmon trout; out of this lake comes the Sevier River, which flows north into the Sevier Lake.

Immediately in the vicinity of the city, is an extensive bituminous coal mine.

Iron ore of superior quality, eighty per cent. pure iron, is found in great quantities; four miles from the city are two mountains of solid ore.

Iron works are in successful operation, all the railroad iron necessary to complete a road from there to San Bernandino, can be procured here.

AFFECTING INCIDENT AT CEDAR CITY. 211

This city is destined to become a great place of business, and, in case the Pacific Railroad does not come through or near Great Salt Lake City, it will be the channel through which all importations for the Territory of Utah will come, it being only about four hundred and fifty miles from San Diego, on the Pacific coast; a distance frequently travelled in ten days.

I renewed my acquaintance with the president, Henry Lunt, with much pleasure, I remained at his house during my stay, and to himself and kind lady, (they are among those who deprecate the spiritual wife system), I was indebted for many little attentions and civilities.

Mr. Lunt was about visiting the city of New York on his way to England, and I gave him a letter of introduction to my family, which he delivered afterwards in person, before I arrived at home.

The morning after my arrival, I arose very early, and taking my sketch-book along, I sauntered around the city; in the course of my peregrinations, I saw a man walking up and down before an adobe shanty, apparently much distressed; I approached him, and inquired the cause of his dejection; he told me that his only daughter, aged six years, had died suddenly in the night; he pointed to the door, and I entered the dwelling.

Laid out upon a straw mattrass, scrupulously clean, was one of the most angelic children I ever saw. On its face was a placid smile, and it looked more like the gentle repose of healthful sleep than the everlasting slumber of death.

Beautiful curls clustered around a brow of snowy whiteness. It was easy to perceive that it was a child lately from England, from its peculiar conformation I entered very softly, and did not disturb the afflicted mother, who reclined on the bed, her face buried in the pillow, sobbing as if her heart would break.

Without a second's reflection I commenced making a sketch of the inanimate being before me, and in the course of half-an-hour I had produced an excellent likeness.

A slight movement in the room caused the mother to look around her. She perceived me, and I apologized for my intrusion; and telling her that I was one of the Governor's party who arrived last night, I tore the leaf out of my book and presented it to her, and it is impossible to describe the delight and joy she expressed at its possession. She said I was an angel sent from heaven to comfort her.

She had no likeness of her child.

I bid her place her trust in Him "who giveth and taketh away," and left her indulging in the excitement of joy and sorrow. I went out unperceived by the bereaved father, who was still walking up and down, buried in grief. I continued my walk, contemplating the strange combination of events, which gave this poor woman a single ray of peace for her sorrowing heart.

When I was about starting the next day, I discovered in the wagon a basket filled with eggs, butter, and several loaves of bread, and a note to my address containing these words—"From a grateful heart."

19*th*.—The Governor and a portion of the party proceeded to-day, to the city of Harmony, twenty-two miles farther south.

Parley Pratt and the party with whom I intended to

travel to California, remained behind to complete their outfit of provisions.

At this point, the road to San Bernandino branches out thirty miles to the westward. We shall proceed on our journey, on the return of Brigham Young from Harmony.

The Payides, or Piedes, were considered the most degraded set of Indians in the Territory, living on reptiles, insects, roots, etc., and going about in a state of nudity.

Since the settlement of Cedar City, they have become more civilized; many of them live within the walls of the city. The Mormons have supplied them with clothes, and proper food. The Indians have become of very great assistance in ploughing and reaping. Several acres of ground have been placed under cultivation, and appropriated for the use of the Indians. They are now acquiring the arts of agriculture and husbandry.

A large number of them have been baptized into the Mormon faith.

It is really astonishing to see the sacrifices and personal privations to which these people willingly, and uncomplainingly submit. Hundreds of families who formerly lived more comfortably at home, are now contented with a mud hut, twelve to fifteen feet square, with a single room, in which they cook, eat, and sleep. In some of them I have seen eight persons, including children, yet they are perfectly happy in the plan of salvation held out to them by the religion they have embraced.

21*st*.—The Governor and party arrived this evening from Harmony.

He has appointed the following gentlemen to take up

a permanent residence with Wakara's band of Utahs, viz. :—

Porter Rockwell, James A. Bean, interpreter; John Murdoch, and John Lott. These persons will follow them in their wanderings, and will, most probably, prevent many depredations and murders.

22nd.—Our party intend starting for California, some time during this day. I breakfasted with Gov. Young; he has given me a letter of introduction to the President of San Bernandino, and all Mormons everywhere. He says I have but to show it, and it will procure me all I require, at any time. I have just taken leave of him and his lady, as well as of the rest of the party.

CHAPTER XXXIII.

On the Road to California—Iron Springs—Meadow Springs—Entrance to Las Vegas de Santa Clara—Prairie Flowers—Rim of the Basin—Santa Clara River—Difficulty of Crossing with Wagons—Wounded Indian—Serpentine Course of the River—Waterfall —Natural Cave.

At three o'clock, our party, consisting of twenty-three Mormons, missionaries to the Sandwich Islands, under command of Parley Pratt, started on their journey. We have six wagons and teams. A woman who is going to her husband at San Bernandino, has permission to accompany us. She also has a wagon and team, but her horses look as if they would not travel fifty miles. She is an encumbrance, and I anticipate trouble with her. We proceeded twelve miles, and camped at Iron Springs, with good water.

22d.—At seven this morning, we were on our road, travelling due west, until two o'clock, when we camped on Penter Creek, twenty-five miles distant from last camp.

The road now forms an elbow, and heads to the south. We followed the course, until we came to Meadow Springs, the entrance to Las Vegas de Santa Clara, noted on Fremont's map—distance twelve miles from noon camp.

This stream is clear and cool. The meadows abound in good grass and rushes, while the surrounding mountains would afford sustenance to thousands of cattle and sheep.

23*d*.—The weather last night was cool and delightful. This morning we left camp at half past seven o'clock, and followed the road in the centre of the valley meadow, to the base of a picturesque mountain, studded with large cedars and umbrageous foliage.

The meadow formed a perfect carpet of various colored flowers, among which were larkspurs, lupines, and many varieties of wild flowers which I have never before seen. I have gathered and preserved specimens of those I considered most valuable.

The contrast of the colors of prairie flowers, as they are thrown carelessly on nature's carpet, is truly wonderful; the greatest harmony prevails—you see the yellow and purple, green and red, orange and blue, arranged always in juxtaposition, producing the primitive colors of a ray of light, through which medium only we are able to distinguish them.

The ancient masters always produced harmony in their pictures because they closely studied nature; at the same time, they could not have known the science of colors, as there is no work extant on the theory of colors, when Raphael or Titian lived. Modern researches have discovered the reasons why nature is thus harmoniously beautiful in all her varied dresses.

The works of modern artists, therefore, should be always correctly delineated, as they not only have the same nature to study from as the ancients had, but science has assisted them with theoretical problems, founded on scientific investigations, in the different branches of Natural Philosophy.

* * * * * *

The road continued through a romantic pass, which wound around the foot of the mountains.

When we reached the divide where the waters flow towards the Gulf of California, the scene that presented itself was grand and sublime.

We camped on the banks of a beautiful stream, the Santa Clara, on the margins of which I observed the rose-tree, in full bearing, also cottonwood, ash, besides shrubs of different kinds, all in bloom. The air was filled with fragrance, and the scene presented a harmonious and refreshing landscape. This paradise is without a solitary living human inhabitant. These plants and flowers are literally

"Wasting their sweetness on the desert air."

We travelled twenty miles this morning, when, after giving our horses a resting-spell, we continued on our journey through this luxuriantly beautiful valley, crossing and re-crossing the Santa Clara six times. This river runs in a serpentine direction, almost due south, the waters of which were, at this time, much swollen. At the last crossing, my mule went in over his head, and I got a wetting as the price of my ferriage.

The wagons had to be pulled over quickly, with all the horses attached to them, by long ropes; the current was so strong as nearly to overturn them. Almost everything at the bottom of the wagons was wet.

The east side of the river, is a continuation of picturesque, abrupt rocks, very much the appearance of the cañons on Grand River, except that the formation is a black ironstone rock, while that of the Grand River is sandstone.

The Santa Clara River, has no connection with the

Seveir River, as was formerly supposed, but is one of the tributaries of the Great Colorado, emptying into the Gulf of California, while the Seveir River empties into Seveir Lake.

We camped on this romantic stream, and at night I took a refreshing bath in its crystal waters.

24th.—At an early hour this morning, our camp was visited by a number of Paiede Indians; they were almost in a state of nudity; we supplied them with food, and some few clothes. One of them, who walked lame, said, he was shot by an exploring party, about ten years ago—corresponding with Col. Fremont's first expedition over this country. With those Indians Col. Fremont had several skirmishes, and I have no doubt, he was wounded in attempting to waylay that expedition. One of the men told him, I was an American, in contradistinction to Mormon. "Ha!" said he, pointing to his wound, "I got that from Mericats"—he looked very savagely at me, and I have no doubt, would have taken delight in making me a target for his arrows: if I had told him I was one of Col. Fremont's men, I am pretty sure I would have had to give him satisfaction. This man followed our camp on foot several days afterwards.

We left camp at eight o'clock, our road lay through scenery similar to that presented yesterday. We crossed the Santa Clara, six times to-day, making twelve crossings, in as many miles. Box, elder, cottonwood, honey locust, grow luxuriantly all along the river, about a mile from the end of the valley where we left it. There is a romantic fall of water on this stream. The fall is twelve feet; on the opposite side of the road there is a natural cave, formed in the red sandstone, which overhangs the road, of nearly fifty feet in depth, and thirty feet high. I explored it, and found only the remains of some Indian articles.

It was about this spot where Lamphere was killed a few weeks before, a description of whose murder I gave in my notes of Salt Lake City. We exercised great vigilance while in camp, and also while travelling through the dense undergrowth of many parts of this river. I looked for enemies in every tree, and was truly rejoiced when we reached the open country again.

CHAPTER XXXIV.

Romantic Pass—Rio Virgin Valley—Sterile County—River Bottoms—Acacia Groves—Abrupt Descent—Formation of the Country—Pah Utahs—Indian Bow and Arrows—Orange color Berries—Effect on the System—Digger Indians—Baptized into Mormon Faith—Steep descent—Divide between Rio Virgin and Muddy Rivers—Difficult travelling—Muddy River described—Author lends his Horse—Approach to the "Great Desert."

WE slowly ascended some sloping hills, which brought us after an hour's ride, on the broad table land. The view then back towards the valley, was sublime beyond description. I made a sketch of it on the spot. Continuing our travel for two hours we halted at a spring of clear water, impregnated with iron. We watered our animals, as it was the last water we should see, until we arrived at the Rio Virgin (Virgin River), twenty-five miles distant; an hour was allotted for the animals to crop some (of the anomaly of this country), bunch grass, which abounded near the spring. We then started for the Rio Virgin, the approach to which, was through the most beautiful and romantic pass I ever saw: it is a natural gorge, in a very high range of mountains of red sandstone, which assume, on either side, the most fantastic and fearful forms; many look as if they were in the very act of falling on the road below them.

The valley of this pass is narrow, but abounds in the most luxuriant grasses and delicate-tinted flowers; a flowering shrub, growing to the height of fifteen feet, exhaling delightful perfume, abounds along the road. Pines and cedars start out from among the rocks, on the sides of the pass, towering one above the other, like Ossa upon Pelion. I have travelled through the beautiful passes in the Rocky and Warsatch Mountains, but I have seen nothing that could excel this, either for the facilities of a railroad, which could be constructed through it without grading, or for the magnificence of the combinations which are requisite to produce effect in a grand landscape. This pass is about six miles through.

Suddenly, as you are about to emerge from this pass, through the opening of the mountains, I beheld the valley of the Rio Virgin at sunset, bursting upon me in all the glory and sublimity of a perfect picture. The view in the distance is unbroken for many miles; generally the scene is blocked in by mountains at short distances.

We descended gently into an extensive valley, sterile to a degree, which seemed to be peculiarly adapted to the growth of a species of palm, called in the West Indies the Spanish needle; this and a dwarf species of artemisia, was the only vegetation visible. The soil is sandy, and embroidered as it were, artificially, with parterres of small pebble stones, arranged with amazing regularity, for many miles, over which our wagons rattled, and bounced amusingly enough to those, who preferred a ride on horseback, to a seat in them. At eight o'clock in the evening, we camped on the banks of the Rio Virgin, the waters of which were also very high. I expect great difficulty in crossing with the wagons to-morrow. Thermometer at noon 90.°

25th.—We left camp this morning, at half-past seven

o'clock. Our road led over a sandy bluff, which was most tiresome to our animals. After a stretch of three miles, we abruptly descended some two hundred feet into the bed of the river, which we crossed with much difficulty, as the water was over the bottoms of the wagons.

The road led through a continuous grove of acacias (*spirolobeum odoratum*), in full bloom, interspersed with a few cottonwoods. We found this road, also, to assume a serpentine course, which created the necessity to recross it seven times, by noon camp.

I noticed on this river a beautiful tree, covered with white flowers hanging in tassels like the flowers of the locust; it resembles the willow, with its long narrow leaves. It is about as large as the weeping willow; it is, certainly, the most beautiful ornamental tree I ever saw.

There are two species of acacias, one closely resembling the opoponax, the other bearing long white blossoms and spiral seed vessels.

These trees abound with doves, which, with the mocking-bird, are the only kinds of the feathered tribe I noticed.

The formation on both sides of the river is a conglomerate, or pudding-stone, with layers of sandstone.

Thousands of party-colored flowers cover the dry, sandy bottoms. It seems a marvel to me how the loose dry sand can yield nourishment sufficient to enable them to grow so luxuriantly.

We travelled twenty miles to-day, along the river, and camped at six o'clock on the road, with good bunch grass on the hills around.

A number of Pahutes came into camp this evening;

they were friendly, and also hungry. We gave them supper. I procured from one of them a bow, made of a single horn of the big horn sheep, covered on the outside with deer-sinew, which they chew until it forms the consistency of thick glue; they then cover the back of the bow with it to increase its strength. I also procured from them a quiver full of steel and obsedian pointed arrows, in exchange for some articles of clothing.

26th.—We left camp this morning at eight o'clock; our road lay through a complete forest of bushes about three feet high, covered with an orange-colored berry. The Indians, who followed our camp, said they were good to eat.

Nearly all of the party partook of them, as they tasted well. A short time after eating them I fell sick, and they affected me in the same manner as if I had taken an emetic. All the camp were affected in the same manner. No other unpleasant consequences followed our imprudence.

The scenery around is uninteresting. We camped at noon, for luncheon, after having crossed the river five times to-day. The sun is very hot, and riding exposed to its influence is not very pleasant.

After resting our animals and satisfying the inner man, we resumed our journey, and camped on the river, having crossed and recrossed it fifteen times.

The high bluffs immediately over our camp, are covered with Indians, all armed. I hardly think they will have the temerity to attack us. We travelled to-day twenty miles.

The most degraded and lowest in the scale of human beings are the Digger, or Piede Indians, of the Rio Virgin and Santa Clara Rivers. Our camps were frequently visited by them. I have often observed them with lizards, and snakes, frogs and other reptiles, strung on a stick over their shoulders, endeavoring to sell or trade for articles of clothing. At certain seasons they dig for roots to subsist on. They go about perfectly nude, with the exception, sometimes only, of a piece of deer-skin around their loins. They are expert thieves, and great vigilance must be used to prevent them from robbing you before your very eyes.

The Indians on the Muddy River are a little higher in the scale of civilization. At one of their villages at which I rested, I found corn and wheat under excellent cultivation, the women grinding it between stones. This improved state is owing to the Mormons, who travel continually on this route to and from San Bernandino. From them they obtained the seed, and several implements of agriculture. The chief and half-dozen others in this village had been baptized in the Mormon faith. The Mormons have acquired the Piede language, and have collected many of the words and sentences, which they have printed.

The following is an illustration of a few sentences arranged in the Piede dialect:

Cot-tam-soog-away,	I don't understand.
Huck-ku-bah-pe-qua?	Where are you going?
Im-po-pe-shog-er,	What are you hunting?
Cot-tam-nunk-i,	I don't hear.
Koot-sen-pungo-pe-shog-er,	I am hunting cattle.
Huck-ku-bah-pah?	Where is the water?
Pah-mah-ber-karry,	The water is over yonder.
Topets-karry,	There is a spring there.
Huck-ku-bah-kah-bah-poni-koe,	Where did you see the horse?
Kah-ponikee-kan-e-gab,	I saw the horse at the foot of the mountain.

NUMBERS.

Soos,	1
We-ioone,	2
Pi-oone,	3
Wol-soo-ing,	4
Shoo-min,	5
Nav-i,	6
Nav-i-ka-vah,	7
Nan-ne-et-soo-in,	8
Shoo-koot-spenker-mi,	9
Tom-shoo-in,	10
Wam-shoo-in,	20
Pi-oone-shoo-in,	30
Wol-so-i-mi-shoo-in,	40
Shoo-mo-mo-shoo-in,	50
Nav-i-me-shoo-in-ny,	60
Nav-i-kah-mi-sho-in,	70
Nan-ne-et-soo-e-mi-shoo-in,	80
Shu-cut-spinker-mi-shoo-in,	90
Wah-kut-spinker-mi-shog,	100

27th.—At eight o'clock to-day, we were on the road, which turned towards very high bluffs. We found the ascent so steep that it was necessary to unharness all the horses from the wagons, and attach them all to one wagon, making fourteen animals dragging one vehicle up this difficult eminence—the men also assisted. This ascent was about 400 yards, and an angle of 35 degrees. We were busily occupied three hours, in taking all our wagons to the table land above. Our course then lay over a barren desert, due west.

The road was covered with a loose fossilliferous rock, very flinty, and painful for our animals to travel. We travelled over the same character of road for twenty miles, and then descended into the valley of the Muddy River, through a deep, irregular cañon of at least three miles in length. We reached the river at five o'clock, after a toilsome and most disagreeable day's travel.

We found excellent grass for our animals on its banks; the temperature was 90° Fahrenheit, which is not much above the average of the coldest weather. This river, supposed to be the Rio de los Angeles, vulgarly called Muddy, takes its rise from hot springs in the mountains. The Indian name is "Moap." The Indian name of the Santa Clara is "Tonequint"—Rio Virgin, is "Paroos." The water is clear and pleasant to the taste, and by no means deserves the name of Muddy.

As soon as my mule was unsaddled, I was in the water, and enjoyed a delightful bath, which was refreshing after such a long hot ride.

We intend to encamp here for a day, to recruit our animals, and make some little preparation for our travel over the dreaded Jornada, a distance of fifty-five miles, without a drop of water or a blade of grass for the animals. Jornada means a journey, viz.: a journey on which you cannot stop; for your animals, if they rested without food or water for such a distance, would go mad; therefore, it is necessary to continue, and push right through, on one stretch, for fifty-five miles. It is most serious to contemplate, but "no hay remedia."

My mule is in good order, and I trust to him to carry me safely over it. Yesterday I found it necessary to lend my horse to the woman who accompanies us; one of her horses gave out, and my horse was the only spare animal. It is just what I expected; but as she is along with us, we must assist her at all hazards. The camp is filled with Diggers; Fremont calls them Pah Utahs, *i. e.*, Utahs living on the water.

These Indians, we find are great thieves; they appear friendly, and we put up with their peccadilloes for policy's sake.

CHAPTER XXXV.

Preparations to Cross the Jornada—Fifty-five Miles without Water or Grass—Deserted Wagons on the Road—Dead Oxen and Mules—Emigrant Party—Clouds of Dust—Oasis—Delicious Water—Extraordinary Fresh Water Buoyant Spring—Impossibility for a Man to sink in it—Never before Described—Another Jornada of Forty Miles—Col. Reese's Train—Detention—Reese Cut off—Snow-Capped Mountains—Bad Roads—Mineral Springs—My Mule in Harness—Animals giving out.

28*th*.—At about three o'clock, the order was given to fill up the water cans, as we were about to traverse this immense desert where water was not to be had; every vessel that could possibly be used, was immediately put in requisition—canteens, kegs, bottles, cans, etc.

At four o'clock, having harnessed up the horses, and saddled my mule, we were on the road, which led through a loose stony ravine, with much sand; it was very heavy travelling, and our animals moved through it with a great deal of difficulty.

We travelled thus for eleven miles, and then gradually ascended the table land, on a harder and better road.

We commenced our journey in the afternoon, that we might have the benefit of the night air to travel in; a cool, north wind tempered the atmosphere, and we continued the journey through this sterile, bare, and uncovered country, until midnight, when we halted and refreshed our animals with water from our reservoirs. After a rest of three hours, we resumed our journey, and at ten o'clock in the morning of the 29th, we had crossed this dreaded Jornada without any accident, and camped on a narrow stream of deliciously cool water, which distributes itself about half a mile further down, in a verdant meadow bottom, covered with good grass.

This camp ground is called by the Mexicans, Las Vegas. Once more, we had plenty of grass for our fatigued animals, and we determined to rest here, during the day and night.

We passed a number of deserted wagons on the road; chairs, tables, bedsteads, and every article of housekeeping, were strewn along our path. The emigrant party who had preceded us about ten days, from Parowan, to lighten their wagons, threw out first one article and then another, until everything they had, was left on the road. It was not difficult to follow their trail; in one hour I counted the putrid carcasses of nineteen oxen, cows, mules and horses; what a lesson to those who travel over such a country, unadvised and unprepared.

A strong north wind blew during the morning, which raised clouds of dust, completely and unresistingly filling our eyes with a fine white dust, although I used goggles to prevent it.

The delightful and refreshing water of this oasis, soon purified me, and now, having crossed the desert, bathed and breakfasted, I feel more comfortable, both mentally and physically.

Mezquite, (alga robia) are the only trees growing near this stream.

30*th*.—We remained at camp all day yesterday, and left this morning at ten o'clock.

We followed up this delicious stream for about three miles; I was curious to see from whence it flowed, the general character of the country indicating that we were not far from its source. Several of us turned from the road, and at a short distance, we found its head waters. It was a large spring, the water bubbled up as if gas were escaping, acacias in full bloom, almost entirely surrounded it—its was forty-five feet in diameter; we approached through an opening, and found it to contain the clearest and purest water I ever tasted; the bottom, which consisted of white sand, did not seem to be more than two feet from the surface.

Parley Pratt prepared himself for a bathe, while I was considering whether I should go in, I heard Mr. Pratt calling out that he could not sink, the water was so bouyant. Hardly believing it possible that a man could not sink in fresh water, I undressed and jumped in.

What were my delight and astonishment, to find all my efforts to sink were futile. I raised my body out of the water, and suddenly lowered myself, but I bounced upwards as if I had struck a springing-board. I walked about in the water up to my arm-pits, just the same as if I had been walking on dry land.

The water, instead of being two feet deep, was over fifteen, the depth of the longest tent pole we had with us. It is positively impossible for a man to sink over his head in it; the sand on its banks was fine and white. The temperature of the water was 78°, the atmosphere 85°.

I can form no idea as to the cause of this great phenomenon; Col. Fremont made observations on the spot in 1845, and marked its existence on his map as Las Vegas; but he has since told me he did not know of its bouyant qualities, as he did not bathe in it. In the absence of any other name, I have called it the Buoyant Spring.

Great Salt Lake possesses this quality in a great degree, but that water is saturated with salt; this is deliciously sweet water; probably some of the *savans* can explain the cause of its peculiar properties. We lingered in the spring fifteen minutes. Twenty-three men were at one time bobbing up and down in it endeavoring to sink, without success. I made drawings of this spot, and the surrounding mountains.

If it were not for this " blessed water," it would be almost impossible for man to travel across these deserts; the next water is at Cottonwood Springs, twenty miles distant.

Twenty miles S. S. W. of us, is a high range of mountains; the two centre ones were covered with snow.

We travelled through them by a romantic pass; the road was level although heavy, being composed of small pebbles, and loose sandstone. I perceived no vegetation, but the usual desert shrubs. In the bosom of these mountains we came to a spring of clear cold water, near which grew luxuriantly, cottonwood, acacias, and a kind of willow in full bloom. We encamped on tolerably good grass.

We have before us another Jornada of forty miles for to-morrow's work.

I collected from the acacias about an ounce of good "gum arabic." I think it is to all appearance the same tree which produces it in the West Indies.

31*st*.—We made an early start this morning, and commenced ascending to a high pass, in a rocky range of lofty mountains, studded with pine, and cedars; the road was very heavy, with loose cobble-stones, and sand. The ascent occupied four hours. We halted at about a mile on the other side, and found a spring of good water.

We met encamped here, Colonel Reese's train, from San Bernandino, bound for Great Salt Lake City. They were in a most distressed state. They had lost a great many of their animals on the desert, and were unable to proceed with the whole expedition. Their wagons were loaded with necessaries and merchandise for the settlements; they had to send to Cedar City for fresh animals to enable them to continue.

I purchased a small quantity of sugar and tea from them, for which I paid a high price—fifty cents per lb. for brown sugar.

We gave our animals a good rest, and started for the Jornada by a new cut off, discovered by Col. Reese.

We travelled over most uncomfortable roads, the soil, instead of sand as heretofore, is an impalpable white powder, very much like pulverized limestone, sown with large rocks; my eyes, although protected with a vail and goggles, suffered very much the whole way. The old road was south south east, this *cut off* led south south west. It is said, by this route, forty miles of travel is saved, and you escape the salt and bitter springs.

The country is an extensive barren waste, we continued on it until midnight, without finding a blade of grass. We camped until four o'clock, A. M.

June 1st.—We started at day dawn, and have, by our calculation, travelled over forty miles. The snow capped mountains, observed on the 30th, as bearing S. S. W. now bear directly north.

At three o'clock, we camped at a spring, at the foot of a range of high hills of pudding-stone.

The last twenty miles of this day's work, has had a decidedly bad effect on our animals. My mule has been in harness yesterday and to-day, to assist the Mormon lady. One might, as it were, see the flesh go off his body—he has lost at least thirty pounds in the last forty-eight hours. One of our horses gave out, and was shot on the road, a wagon also broke down and was left on the road.

On examining the spring, I found it to be strongly impregnated with sulphur and iron; it is a very pleasant mineral water, although very warm; the thermometer indicated a temperature of 90,° while, when exposed to the atmosphere, it sunk to 65°—at six o'clock, P. M.

2d.—Our road, during the last twenty miles, lay along the dry bed of a creek, until we came to a high range of volcanic rock, where we pushed our way through an intricate pass to the spring which is on the road, immediately after emerging from the cañon.

The ground on which the spring is situated, is rather elevated, the earth is elastic to the tread, and almost any where near it, you can get water by digging eighteen inches. This water is also slightly impregnated with iron.

CHAPTER XXXVI.

Peg-leg Smith—Gold Explorers—Enter upon the Desert—Road strewn with Dead Oxen—Poisoned Atmosphere—Deserted Wagons and Horses—Howling Wilderness—Excessive Heat—Bitter Springs—Polluted by Dead Animals—Bunch Grass—Reflections—Mohahve River—Deserts Surmounted—Horses give Out—On Foot—Dig for Water in the Sand—Pleasant Weather—Snowy Mountains—Crossing of the Mohahve River—Agave Americana—Cajon pass Sierra Nevada—Descent into the Valley of San Bernandino—Arrival at San Bernandino—Variations of the Compass.

WHILE encamped on this spot we met a party of gold explorers from Los Angeles. They had been down on the Colorado, looking for gold, but had been unsuccessful. They were under the command of a man with one leg, known as "Peg-leg Smith," a celebrated mountaineer.

He told me he had been several times across the continent, and had been in this part of the world for some years.

He says he crossed the Rocky Mountains in 1824—30 years ago. He is a weather-beaten old chap, and tells some improbable tales. They are on their way back, and will travel with us; they comprise ten men, all mounted on mules.

To-day two more of our horses gave out; one of them belongs to the wagon which contained my baggage. Mr. Peg-leg Smith tells me these are called Kingstone Springs. I made drawings of the mountains which are near them; they are curiously formed land marks, and may be useful to future travellers. We have another terrible Jornada to pass, a distance of fifty miles.

I hardly think we shall get over it without leaving some of our animals.

At 3 o'clock we started; our course was south west, over a new country. Reese's train was the first who had ventured; none of our party had ever been over, and *I never want to traverse it again.*

In travelling over the vast prairies and mountains it is well that the range of our vision has certain limits. If we could take within scope of our sight, the whole extent of the distance to be travelled, we should most probably give up the original intention as one of the imposibilities; a wise Providence has ordained otherwise. The distance is bounded frequently by high ranges of mountains, which cut off the perspective, or the atmosphere between the eye and the object produces an aërial effect, which obscures like a curtain, the far spread waste, inspiring the wearied traveller with fresh and renewed energy.

"So doth the untrod distance still delude us."

This was decidedly the worst ground I had ever travelled. After 20 miles ride, I saw in the distance, what I took to be a lake, and none of the party knew better. It was an extensive bed of pure white sand, probably fifteen miles in diameter, and may have been once the bed of a lake. Our road lay directly over it, and we proceeded slowly, and with much difficulty; at midnight we rested our animals.

3rd. At 4 o'clock we were on the road again. Carcasses of dead horses and oxen, strewed the way. Some were left to die, and others still warm, although dead. In the space of one mile I counted 40 dead oxen and cows; the air was foully impregnated with the effluvia arising from them. We also passed six deserted wagons, chairs, tables, and feather beds which were left on the road in greater quantities than on the first desert.

At noon we arrived at Bitter Springs, the grounds about which are strewn with dead animals, and the polluted atmosphere at this time, one o'clock, P. M., ranges at 95° in the shade of our wagons, and is nearly unbearable.

This is a howling, barren wilderness; not a single tree or shrub for the last fifty miles, nor is there one in sight now. I did not observe during the last day's travel, a lizard or any sign of animal or insect life. There was plenty of food for wolves, but they dare not venture so far from water.

These springs are not bitter, but possess a brackish taste. There are small springs in different places; the largest admitted one horse at a time to drink, the rest would have to wait until the water was replenished from the earth.

While I write of the sterile and barren desert, over which I have travelled, I cannot but contemplate with admiration the goodness of the Almighty, in placing at intervals, food and water for the sustenance of our animals.

Along the whole road there is not a blade of grass for a distance of fifty miles; but in the immediate vicinity of this spring there are hundreds of acres of the best quality of bunch grass; there is, apparently, the same sandy barren soil, not deriving any nourishment from the spring, which is a mile away.

Without the watchful care of Divine Providence, man would be unable successfully to traverse these deserts.

June 4th.—We left camp at 5½ P. M., and camped at

8 o'clock this morning: 5th, encamped on the Mohahve River. We made 31 miles since last evening.

I return grateful thanks to the Omnipotent for conducting me safely over the mountains of snow, and the dangers of the desert wilderness.

We may now consider the real perils of the journey past. San Bernandino is ninety miles S. W. of us. In four days, I trust we shall arrive in good health and condition.

Yesterday two horses gave out. Our Mormon lady is the sub-tenant of one of our wagons; her own was so heavy as to wear out the animals, she was obliged to leave it on the road. My poor mule is only a shadow of himself, I walked about fifteen miles yesterday, to relieve him. He has now good grass for his supper.

When we struck the Mohahve River, it appeared to be only a dry bed of sand, with a few pools of water about six inches deep. We were very grateful that we found any at all, as our animals were suffering very much for the want of it.

Cottonwood and willows grow abundantly near the banks. The sight of vegetation is refreshing, and indicates our approach to a country more adapted for the purposes of man.

We left camp at four o'clock, in hopes of finding a better camp-ground.

We travelled thirteen miles through loose deep sand, when, turning again to the river, we found a large sluggish pool of water, twenty-five feet in diameter, and one foot deep in the bed of the river, which sinks and rises in the sand for many miles.

Good bunch grass was here in abundance, and our animals are faring sumptuously.

The flowering willow (a dwarf), is the only tree now visible. Thermometer, at day-light, 60°. A strong gale of wind is blowing from the north.

We have been highly favored with pleasant weather during our journey across these deserts, with the exception of a few hours at mid-day: the temperature has been delightful, quite opposite to what I had anticipated.

6th.—We left camp this morning, and continued along the dry bed of the Mohahve River for fifteen miles, when we halted. We dug holes in the sand, and found good pure water.

Our camp-ground is surrounded with fine large cottonwoods, and plenty of bunch grass on the benches near.

7th.—We were on the road at an early hour this morning. We struck across a sandy desert, of about ten miles, and approached the river again, but found no water. We continued along, and at noon halted about five miles further up, with clover, grass, and water in a little pool on the road.

The thermometer at daylight this morning, was down to 40°. Large fires were very comfortable. In the last forty-eight hours, there has been a variation of 60° of the thermometer, in the shade.

The weather is more like October than June.

Two high snowy mountains, bearing S. S. W., almost immediately on our course, indicate our approach to the Nevada Mountains.

At five o'clock, we encamped within five miles of the crossing of the Mohahve River. Abundance of good red clover, grass and plenty of water.

We travelled thirty miles this day.

8th.—At daylight this morning our camp was in active preparation for departure. The temperature

55°, and delightful weather. After an early breakfast, we rode through a beautiful grove of cottonwood, with willow undergrowth. Rose trees in full bloom, with hundreds of other beautiful flowers. This is a fairy land, indeed. What a contrast to the desert of a few hours ago! Grape vines hang gracefully from the branches of lofty trees, while the air resounds with the songs of birds. I noticed numbers of doves, a species of quail with a top-knot (the California quail), herons, and ducks in great numbers on the river.

We crossed the river, which at this place was a running stream, about two hundred yards wide, and fringed with cottonwood and willow trees. After leaving the river, we commenced to ascend gradually to another desert, of seventeen miles. The last five miles was through a forest of muskale (*Agave Americana*), which grow to an immense size; some as large as the greatest oak tree I ever saw. This is a curious tree, the trunk is cylindrical, as if it were turned; its limbs are leafless, except at their extremities, on which grow long narrow leaves, with a sharp prickle at the end. These trees assume the most fantastic forms. At noon we arrived at the summit of the Cajon Pass, in the Sierra Nevada the descent from which is on a saddle or spur of the mountain, on an angle of thirty-five degrees, and the length of the descent is a quarter of a mile, then it becomes more gradual for a mile, until you reach the valley below.

The view from the top of the pass, is grand beyond description—from it, you can see the San Bernandino Mountains, and numberless valleys; from this eminence the Tulara Pass is in view.

The descent of our wagons occupied considerable

time; the team was in front, but the whole force of the men were attached to long ropes at the end of the wagon, to prevent its too rapid descent; the surface of this saddle is perfectly smooth, and a good team of horses easily draws up a wagon over it to the top. There would be no difficulty for two steam engines to propel a train of cars up this natural inclined plain, while the road from Great Salt Lake to San Bernandino, eight hundred and fifty miles, could be laid without any grading; the passes through the mountains being perfectly level, and well adapted for railroad purposes—while the deserts are almost perfect plains.

After descending into the valley, the road to San Bernandino leads through a wide level cañon, in which grow spontaneously abundance of wild oats. We encamped, after journeying ten miles through it, with good water and grass. We travelled thirty-two miles this day.

9th. This morning at daylight the thermometer was at 35°. We left camp early, and continued through the cañon, which was well timbered for twelve miles, we then emerged into the San Bernandino Valley, and at one o'clock, P. M. we all arrived safely at San Bernandino. I collected and preserved numerous specimens of wild flowers, which are yet unclassified.

My mule is in tolerable condition, the last few camps where good clover and grass were obtained, improved him greatly. The horses have all come in very poor, and many of them lame and broken down.

I was kindly received by Gen. Rich, the president of San Bernandino, who showed me many civilities.

San Bernandino Valley, is a tract of most fertile country; it was the seat of a Catholic Mission some years before, but recently purchased by the Mormons for a settlement.

San Bernandino City, contains about one thousand inhabitants, the church owns saw-mills and flouring mills, it is a great agricultural country. Being desirous of reaching the sea-board, I only remained three days here. I mounted my trusty mule, and rode into Los Angeles in twelve hours, a distance of forty-five miles, pretty well for an animal that had just come off the deserts.

Variation of compass:—
at Great Salt Lake City,	15′ 20″ W.
" Fillmore City,	16′ "
" Parowan,	19′ 40″ "
" San Bernandino,	13′ 40″ "
" Los Angeles,	13′ 30″ "

Immediately in the vicinity of Parowan, there are several mountains containing magnetic iron, which accounts for the great variation in that place.

CHAPTER XXXVII.

Journey to Los Angeles—Catholic Missions—Fields of Mustard—California Ladies—Morals of the People—Gamblers—Description of a "Hell"—Climate of Los Angeles—Delicious Fruit—California Wine—Don Manuel Domingues—Rancho—Menada—Breaking a Horse—Portraits of Domingues—Salt Lake—Asphaltum Lake—Hot Springs of San Juar de Campestrano—Analysis—Geological Examination—Remains of a Mastodon—Don Pio Pico—Ground Squirrels—Strychnine—Brothers Labatt—Their Example worthy to be Imitated.

FIELDS OF MUSTARD.

From San Bernandino to Los Angeles, a distance of forty-five miles, the road lay over one continuous field of wild mustard, covering the whole breadth of the valley of Los Angeles, and extending far up into the mountains; it was ripe at the time I travelled through. Millions of acres producing many thousands of bushels, annually go to waste. If coal is ever found in this country, a mustard mill could be profitably worked. At present there is no water power to turn a mill, or fuel to propel an engine for steam works.

CATHOLIC MISSION—LOS ANGELES.

To-day I met Mr. Hildreth, one of the brothers who commanded a large emigrant party, and whose unprovoked and fatal attack upon the Parvain Indians, near

Fillmore, caused that tribe to murder Capt. Gunnison and officers, a description of which I have already given. Mr. Hildreth says that his brother (the commander), and himself had left camp to hunt, and when they returned they were informed of the unfortunate and premature attack of some of his people upon the Indians. It seems that a small number of Parvain Indians came into camp armed with bows and arrows, begging food and clothing at sundown. They were ordered out of camp, they refused. They were told if they gave up their bows and arrows they might remain, and one of the men used force to obtain the bow from an Indian. In the scuffle the American was wounded, whereupon, without any further provocation, a number of rifles were discharged at the Indians, killing several, among whom was an old chief. Capt. Hildreth at once raised camp and proceeded on his journey for fear of the consequences. This fatal event would not have occurred if Capt. Hildreth had been in camp, and he lamented the occurrence.

The California ladies are generally brunettes; some of them with whom I became acquainted were most beautiful and accomplished. Bonnets are unknown. During the morning their magnificent tresses are allowed to hang at full length down their backs. I have seen suits of hair at least three feet long, waving gracefully around a well-formed neck. In the evening a great deal of care and pains are taken to curl and plait it. When they go out, a simple mantilla of black satin or silk, sometimes of colored silk, is gracefully thrown over their heads; they invariably carry a large fan. The most costly material is used for dresses, and the richest and most expensive shawls may be seen worn by the ladies in Los Angeles. Society is very select among the better classes, although there are but few American families residing there.

Alas! for the morals of the people at large; it was the usual salutation in the morning, "Well, how many murders were committed last night?"—"Only four—three Indians and a Mexican." Sometimes three, often two, but almost every night while I was there, one murder, at least, was committed. It became dangerous to walk abroad after night. A large number of American gamblers frequented the principal hotels, and induced the Californians to risk their money at all the famous games of monte, roulette, poker, faro, etc.

When I arrived at San Francisco, I had the curiosity to enter one of the most frequented "hells," to see the process of winning and loosing money. The building selected by the gentleman who accompanied me, was a celebrated one in Clay street. An orchestra of thirty-five musicians, were performing fashionable operatic airs; following the sound, we were introduced into the saloon, which was brilliantly illuminated; it was truly an imposing sight. There must have been over fifty tables, at which presided most beautiful women, dealing out cards, or whirling around a roulette table; at some might have been seen old gentlemen with white hair, to all appearance respectable, and whose proper place seemed to me, to be a magistrate's bench, or a judge's forum. Few or no words are spoken at the table; men silently place their gold on a card, and before a second expires, it is swept away; once out of many times, it is doubled by the player; it remains and he wins: a second time fortune favors, it doubles again; the insatiate vice of selfishness, not satisfied with eight times what he originally staked, leaves his pile, building castles in the air with the imaginary proceeds of his winnings—when in the twinkle of an eye, a gentle sweep from the smiling syren, dissipates his dreams of fortune, and he retires from the hell penniless in reality. Hundreds of men who have acquired by hard work and industry, a little fortune at the mines, and come to town to purchase a bill of exchange to send to their families, are induced to visit one of these places, and in an hour he has lost the labor of months, leaving his family anxiously awaiting remittances which they are doomed never to receive.

These native Californians have been known to borrow money at the enormous rate of six per cent. a month, compound interest, and give their ranchos as collaterals, on purpose to gamble with; many who once were rich, are now reduced to beggary from this cause; the compound interest accumulating so fast, that unable to meet it, the mortgage is foreclosed, and a valuable property sacrificed to the usurious practices of those who call themselves men, for one twentieth part of its real value.

The climate is delightful. The pine-apple, grapes, figs and oranges of the tropics, grow alongside of the pears, peaches and apples, of the temperate regions. The most delicious grapes I ever tasted, are cultivated in large quantities in Los Angeles. Hundreds of tons are annually shipped to San Francisco; peaches, delicious pears, etc., and, in fact, the fruit is cultivated purposely to ship. It yields a good profit and a large income. The vineyards are set out in drills six feet apart, each vine is trained to an upright position supported by rods, until they acquire age. The usual price for grapes was three dollars a hundred pounds as they are on the vines, to be plucked and boxed at the expense of the purchaser, other fruit is also sold by the pound on the tree. Many proprietors have permanent engagements with San Francisco merchants, to sell annually the produce of their vineyards and orchards.

Wine of a superior kind is made in Los Angeles, it is white and dry like the Hockhiemer or Rhenish. A superior article is worth twenty-five dollars for eighteen gallons.

Don Manuel Domingues, a noble specimen of a Spanish gentleman, owns a very large tract of land in Los Angeles county. The San Gabriel, and Los Angeles Rivers run through it, making the property very valuable. It adjoins the large rancho of Mr. Stearns. It was confirmed by the United States government during my short residence at his hospitable mansion, and I painted a large portrait of him to celebrate the event, with the letters patent of his property in his hand. I was prostrated at this gentleman's house by a severe attack of brain fever, superinduced by exposure in travelling over the hot deserts of sand, between Salt Lake and San Bernandino. His good, kind-hearted wife, Donna Gracia, paid me all the attentions and devotion of a mother. For ten days I was delirious, during that time she hardly left my bedside. Doctor Brinkerhoff who resided with them, volunteered his medical advice. To their combined skill and care I owe my final recovery.

I was taken ill the very day I got out to their rancho. If I had been ill in Los Angeles, where I had been residing previously, I should have died for want of attentions which money could not have procured.

I also painted the portraits of Donna Gracia, and one of her daughters.

Don Manuel has several brothers, living at short distances from each other; they have all large families of grown sons and daughters, who meet alternately at each other's houses, when music and dancing is indulged in with unalloyed pleasure. Young gentlemen from town often drive out to spend an evening, and the four weeks I spent there, speaking Spanish and dancing with the beautiful senoritas, conduced much to restore me to the habits of civilized life, which a voyage of nine months, across the continent had almost made me forget.

Dan Manuel has an immense number of oxen, sheep and horses. His menada is said to contain the finest riding animals in California; and it is only by great persuasion that he will sell a choice horse. While I was there, I saw the process of breaking a horse to the saddle. A native Californian lassoes the animal he intends to break, and brings him out of the menada. One end of the lasso he ties around the nose of the horse; a blanket is strapped on his back by a strong surcingle; he then jumps on him, and introducing his knees under the surcingle, he is now firmly seated. On his feet are immense spurs; he touches the horse with them, and off he bounds with the speed of the wind, his rider guiding him with perfect ease. Now he plunges—see him rearing! but his master is on him, and his efforts to dismount him are unavailing. After he is exercised in this manner for an hour, he is turned into pasture, picketed, and not suffered to run with the menada afterwards.

The mares are of comparatively little worth; they are never used as beasts of burthen, or for riding; they are kept for breeding purposes. I have seen a magnificent animal sell for forty dollars, while geldings, not superior in quality, brought two hundred dollars.

On this rancho, towards San Pedro, is a salt lake, which was being worked by a company of gentlemen. The salt is of superior quality, and brings a good price in Los Angeles.

On this same place, near the shores of the Pacific Ocean, there is a lake of bitumen or asphaltum, used almost altogether in Los Angeles, as covering for the roofs of houses. In winter it does very well, but the dropping of hot pitch from the eaves of the houses in hot weather, is not agreeable. Large quantities of it are, in consequence, on the side-walks, which, in warm weather, acts like bird-lime; for if you meet a friend, and stop accidentally on it, there you both are fixed for the moment. Gentlemen's clothing is frequently spoiled by this material. It is highly inflammable; an excellent gas might be obtained from it. I have seen it used on steamboats, to get up steam quickly.

The mission of San Juan de Campestrano is not far from this rancho. Near it are the celebrated hot springs of that name.

For the following analysis, I am indebted to Dr. Wm. P. Reynolds, whom I met on the steamer to San Francisco:—

Sulphur,	40
Nitre,	11
Ammonia,	7.5
Potassa,	9.3
Lime,	7.2
Phosphorus,	6
Iron,	13
Soda,	6
	100

These hot springs of San Juan de Campestrano excel all others in the neighborhood (and there are many), in regard to their medicinal virtues, both from their chemical combinations and the results obtained by their healing qualities in all those diseases for which the chalybeates are reported to cure.

In making geological examinations on Domingues' land, I had the curiosity to dig into a mound of earth raised up several feet from the surface, and not fifty yards from the dwelling-house. I found several pieces of large size petrified bone, too colossal for horses or oxen. Procuring a pick-axe, I penetrated further, and was gratified in exhuming portions of a mastodon. I collected four perfect teeth; the largest weighed six pounds. I destroyed several with my axe, before I realized their value. Portions of the tibia I also got out perfect. These interesting antediluvian relics I took with me to Los Angeles, where I met Mr. Trask, the State geologist of California. At his request I presented two specimens of the teeth to the State Geological Society, the rest Mr. Trask took charge of for me, to deliver in California. I have never seen a report of my present to the society, and when I met Mr. Trask at San Francisco, they had not yet been shipped from Los Angeles. I regret very much that I allowed them to leave my own possession, as I promised one of the teeth to Col. Fremont, and, in consequence, have not been able to fulfill it.

These huge animals are granivorous, and must have consumed trees on the mountains; around Los Angeles there is no sign of a tree, and on the vast plains in the centre of which I found these petrified remains, there is nothing but short grass and mustard. Query, how came the mastodon in the place I found it? did it die there? or was it washed down from the mountains? I leave this interesting investigation to more scientific minds.

At Los Angeles, I painted the portraits of the ex-governor, Don Pio Pico, and several other gentlemen.

The whole country of Southern California, especially in Los Angeles county, is infested with millions of ground squirrels, which destroy vegetation, and are great nuisances to farmers, as well as to the community; they domesticate themselves in houses, and I have seen them jump on the dinner-table, overturning tumblers, etc. The country is overrun with them; various methods have been suggested to destroy them, but without effect; the most successful, however, is strychnine—large quantities of which are imported into California, for this express purpose. This virulent and active poison, for this reason, becomes an important article of trade.

These squirrels, form the principal food of the numerous bands of degraded Indians, who live near the settlements.

To the brothers Samuel and Joseph Labatt, merchants of Los Angeles, I am indebted for many acts of kindness; men who anticipate the necessities of their fellow-man, and spontaneously offer *money advances* to a perfect stranger, I have not often met with, " but when found, I make a note of it."

With the view of not interrupting the incidental part of this book, I have preferred to place at the end of it, several sermons, illustrating the oratorical powers of Brigham Young, and some of his apostles and counsellors, as well as the " Revelations to Joseph Smith, on the patriarchal order of matrimony, or plurality of wives," (presented to the author by his excellency), which is the basis of the spiritual wife system, as now practised by the Mormons.

MORMONISM.

SPIRITUAL WIFE SYSTEM.

A REVELATION ON THE PATRIARCHAL ORDER OF MATRIMONY,
OR PLURALITY OF WIVES.

Given to Joseph Smith, the Seer, in Nauvoo, July 12th, 1843.

1. Verily, thus saith the Lord unto you my servant Joseph, that inasmuch as you have inquired of my hand, to know and understand wherein I, the Lord, justified my servants, Abraham, Isaac, and Jacob; as also Moses, David, and Solomon, my servants, as touching the principle and doctrine of their having many wives and concubines: Behold! and lo, I am the Lord thy God, and will answer thee as touching this matter: Therefore, prepare thy heart to receive and obey the instructions which I am about to give unto you; for all those who have his law revealed unto them, must obey the same; for behold! I reveal unto you a new and everlasting covenant, and if ye abide not that covenant, then are ye damned; for no one can reject this covenant, and be permitted to enter into my glory; for all who will have a blessing at my hands, shall abide the law which was appointed for that blessing, and the conditions thereof, as was instituted from before the foundations of the world: and as pertaining to the new and everlasting covenant, it was instituted for the fullness of my glory; and he that receiveth a fullness thereof, must and shall

abide the law, or he shall be damned, saith the Lord God.

2. And verily I say unto you, that the conditions of this law are these: All covenants, contracts, bonds, obligations, oaths, vows, performances, connections, associations, or expectations, that are not made and entered into, and sealed, by the Holy Spirit of promise, of him who is anointed, both as well for time and for all eternity, and that too most holy, by revelation and commandment, through the medium of mine anointed, whom I have appointed on the earth to hold this power (and I have appointed unto my servant Joseph to hold this power in the last days, and there is never but one on the earth at a time, on whom this power and the keys of the priesthood are conferred), are of no efficacy, virtue, or force, in and after the resurrection from the dead: for all contracts that are not made unto this end, have an end when men are dead.

3. Behold! mine house is a house of order, saith the Lord God, and not a house of confusion. Will I accept of an offering, saith the Lord, that is not made in my name? Or, will I receive at your hands, that which I have not appointed? And will I appoint unto you, saith the Lord, except it be by law, even as I and my Father ordained unto you, before the world was? I am the Lord thy God, and I give unto you this commandment, that no man shall come unto the Father, but by me, or by my word which is my law, saith the Lord; and everything that is in the world, whether it be ordained of men, by thrones, or principalities, or powers, or things of name, whatsoever they may be, that are not by me, or by my word, saith the Lord, shall be thrown down, and shall not remain after men are dead,

neither in nor after the resurrection, saith the Lord your God: for whatsoever things remaineth, are by me; and whatsoever things are not by me, shall be shaken and destroyed.

4. Therefore, if a man marry him a wife in the world, and he marry her not by me, nor by my word; and he covenant with her so long as he is in the world, and she with him, their covenant and marriage is not of force when they are dead, and when they are out of the world; therefore, they are not bound by any law when they are out of the world; therefore, when they are out of the world, they neither marry, nor are given in marriage, but are appointed angels in heaven, which angels are ministering servants, to minister for those who are worthy of a far more, and an exceeding, and an eternal weight of glory; for these angels did not abide by law, therefore they cannot be enlarged, but remain separately and singly, without examination, in their saved condition, to all eternity, and from henceforth are not Gods, but are angels of God for ever and ever.

5. And again, verily I say unto you, if a man marry a wife, and make a covenant with her for time, and for all eternity, if that covenant is not by me, or by my word, which is my law, and is not sealed the Holy Spirit of promise, through him whom I have anointed and appointed unto this power, then it is not valid, neither of force, when they are out of the world, because they are not joined by me, saith the Lord, neither by my word; when they are out of the world, it cannot be received there, because the angels and the Gods are appointed there, by whom they cannot pass; they cannot, therefore, inherit my glory, for my house is a house of order, saith the Lord God.

6. And again, verily I say unto you, if a man marry a wife by my word, which is my law, and by the new and everlasting covenant, and it is sealed unto them by the Holy Spirit of promise, by him who is anointed, unto whom I have appointed this power, and the keys of this priesthood, and it shall be said unto them, ye shall come forth in the first resurrection; and if it be after the first resurrection, in the next resurrection; and shall inherit thrones, kingdoms, principalities, and powers, dominions, all heights and depths, then shall it be written in the Lamb's Book of Life, that he shall commit no murder whereby to shed innocent blood; and if ye abide in my covenant, and commit no murder whereby to shed innocent blood, it shall be done unto them in all things whatsoever my servant hath put upon them, in time, and through all eternity, and shall be of full force when they are out of the world; and they shall pass by the angels, and the Gods, which are set there, to their exaltation and glory in all things, as hath been sealed upon their heads, which glory shall be a fullness and continuation of the seeds for ever and ever.

7. Then shall they be Gods, because they have no end; therefore shall they be from everlasting to everlasting, because they continue; then shall they be above all, because things are subject unto them. Then shall they be gods, because they have all power, and the angels are subject unto them.

8. Verily, verily I say unto you, except ye abide my law, ye cannot attain to this glory; for straight is the gate, and narrow the way that leadeth unto the exaltation and continuation of the lives, and few there be that find it, because ye receive me not in the world neither do ye know me. But if ye receive me in the world, then

shall ye know me, and shall receive your exaltation, that where I am, ye shall be also. This is eternal life, to know the only wise and true God, and Jesus Christ whom he hath sent. I am He. Receive ye, therefore, my law. Broad is the gate, and wide the way that leadeth to the death; and many there are that go in thereat; because they receive me not, neither do they abide in my law.

9. Verily, verily I say unto you, if a man marry a wife according to my word, and they are sealed by the Holy Spirit of promise, according to mine appointment, and he or she shall commit any sin or transgression of the new and everlasting covenant whatever, and all manner of blasphemies, and if they commit no murder, wherein they shed innocent blood,—yet they shall come forth in the first resurrection, and enter into their exaltation, but they shall be destroyed in the flesh, and shall be delivered unto the buffetings of Satan, unto the day of redemption, saith the Lord God.

10. The blasphemy against the Holy Ghost, which shall not be forgiven in the world, nor out of the world, is in that ye commit murder, wherein ye shed innocent blood, and assent unto my death, after ye have received my new and everlasting covenant, saith the Lord God; and he that abideth not this law, can in nowise enter into glory, but shall be damned, saith the Lord.

11. I am the Lord thy God, and will give unto thee the law of my Holy Priesthood, as was ordained by me, and my Father, before the world was. Abraham received all things, whatsoever he received, by revelation and commandment, by my word, saith the Lord, and hath entered into his exaltation, and sitteth upon his throne.

12. Abraham received promises concerning his seed,

and of the fruit of his loins,—from whose loins ye are, viz.: my servant Joseph—which were to continue, so long as they were in the world; and as touching Abraham and his seed, out of the world, they should continue; both in the world and out of the world should they continue as innumerable as the stars; or, if ye were to count the sand upon the sea-shore, ye could not number them. This promise is yours also, because ye are of Abraham, and the promise was made unto Abraham; and by this law are the continuation of the works of my Father, wherein he glorifieth himself. Go ye, therefore, and do the works of Abraham: enter ye into my law, and ye shall be saved. But if ye enter not into my law, ye cannot receive the promises of my Father, which he made unto Abraham.

13. God commanded Abraham, and Sarah gave Hagar to Abraham, to wife. And why did she do it? Because this was the law, and from Hagar sprang many people. This, therefore, was fulfilling, among other things, the promises. Was Abraham therefore, under condemnation? Verily, I say unto you, *Nay;* for I the Lord, commanded it. Abraham was commanded to offer his son Isaac; nevertheless, it was written, Thou shalt not kill. Abraham, however, did not refuse, and it was accounted unto him for righteousness.

14. Abraham received concubines, and they bare him children, and it was accounted unto him for righteousness, because they were given unto him, and he abode in my law: as Isaac also and Jacob did none other things than that which they were commanded; and because they did none other things than that which they were commanded, they have entered into their exaltation, according to the promises and sit upon

thrones, and are not angels, but are Gods. David also received many wives and concubines, as also Solomon, and Moses my servant; as also many others of my servants, from the beginning of creation until this time; and in nothing did they sin, save in those things which they received not of me.

15. David's wives and concubines were given unto him, of me, by the hand of Nathan, my servant, and others of the prophets who had the keys of this power; and in none of these things did he sin against me, save in the case of Uriah and his wife; and therefore, he hath fallen from his exaltation, and received his portion; and he shall not inherit them out of the world; for I gave them unto another, saith the Lord.

16. I am the Lord thy God, and I gave unto thee, my servant Joseph, an appointment, and restore all things; ask what ye will, and it shall be given unto you, according to my word; and as ye have asked concerning adultery, verily, verily I say unto you, if a man receiveth a wife in the new and everlasting covenant, and if she be with another man, and I have not appointed unto her by the holy anointing, she hath committed adultery, and shall be destroyed. If she be not in the new and everlasting covenant, and she be with another man, she has committed adultery; and if her husband be with another woman, and he was under a vow, he hath broken his vow, and hath committed adultery; and if she hath not committed adultery, but is innocent, and hath not broken her vow, and she knoweth it, and I reveal it unto you, my servant Joseph, then shall you have power, by the power of my Holy Priesthood, to take her, and give her unto him that hath not committed adultery, but hath

been faithful, for he shall be made ruler over many; for I have conferred upon you the keys and power of the priesthood, wherein I restore all things, and make known unto you, all things in due time.

17. And verily, verily I say unto you, that whatsoever you seal on earth, shall be sealed in heaven; and whatsoever you bind on earth, in my name, and by my word, saith the Lord, it shall be eternally bound in the heavens; and whatsoever sins you remit on earth, shall be remitted eternally in the heavens; and whosoever sins you retain on earth, shall be retained in heaven.

18. And again, verily I say, whomsoever you bless, I will bless; and whomsoever you curse, I will curse, saith the Lord; for I, the Lord, am thy God.

19. And again, verily I say unto you, my servant Joseph, that whatsoever you give on earth, and to whomsoever you give any one on earth, by my word, and according to my law, it shall be visited with blessings, and not cursings, and with my power, saith the Lord, and shall be without condemnation on earth, and in heaven; for I am the Lord thy God, and will be with thee even unto the end of the world, and through all eternity: for verily I seal upon you your exaltation, and prepare a throne for you in the kingdom of my Father, with Abraham, your father. Behold, I have seen your sacrifices, and will forgive all your sins; I have seen your sacrifices, in obedience to that which I have told you: go, therefore, and I make a way for your escape, as I accepted the offering of Abraham, of his son Isaac.

20. Verily, I say unto you, a commandment I give unto mine handmaid, Emma Smith, your wife, whom I have given unto you, that she stay herself, and partake not of that which I commanded you to offer unto her;

for I did it, saith the Lord, to prove you all, as I did Abraham; and that I might require an offering at your hand, by covenant and sacrifice; and let mine handmaid, Emma Smith, receive all those that have been given unto my servant Joseph, and who are virtuous and pure before me; and those who are not pure, shall be destroyed, saith the Lord God! for I am the Lord thy God, and ye shall obey my voice; and I give unto my servant Joseph, that he shall be made ruler over many things, for he hath been faithful over a few things, and from henceforth I will strengthen him.

21. And I command mine handmaid, Emma Smith, to abide and cleave unto my servant Joseph, and to none else. But if she will not abide this commandment, she shall be destroyed, saith the Lord; for I am the Lord thy God, and will destroy her if she abide not in my law; but if she will not abide this commandment, then shall my servant Joseph do all these things for her, even as he hath said; and I will bless him, and multiply him, and give unto him an hundred fold in this world, of fathers and mothers, brothers and sisters, houses and lands, wives and children, and crowns of eternal lives in the eternal worlds. And again, verily I say, let mine handmaid forgive my servant Joseph his trespasses, and then shall she be forgiven her trespasses, wherein she has trespassed against me; and I the Lord thy God will bless her, and multiply her, and make her to rejoice.

22. And again, I say, let not my servant Joseph put his property out of his hands, lest an enemy come and destroy him, for Satan seeketh to destroy; for I am the Lord thy God, and he is my servant; and behold! and lo, I am with him, as I was with Abraham thy father even unto his exaltation and glory.

23. Now as touching the law of the priesthood, there are many things pertaining thereunto. Verily, if a man be called of my Father, as was Aaron, by mine own voice, and by the voice of him that sent me, and I have endowed him with the keys of the power of this priesthood, if he do anything in my name, and according to my law, and by my word, he will not commit sin, and I will justify him. Let no one, therefore, set on my servant Joseph; for I will justify him; for he shall do the sacrifice which I require at his hands, for his transgressions, saith the Lord your God.

24. And again, as pertaining to the law of the priesthood: If any man espouse a virgin, and desire to espouse another, and the first give her consent; and if he espouse the second, and they are virgins, and have vowed to no other man, then is he justified; he cannot commit adultery, for they are given unto him; for he cannot commit adultery with that that belongeth unto him, and to none else; and if he have ten virgins given unto him by this law, he cannot commit adultery, for they belong to him; and they are given unto him—therefore is he justified. But if one, or either of the ten virgins, after she is espoused, shall be with another man, she has committed adultery, and shall be destroyed; for they are given unto him to multiply and replenish the earth, according to my commandment, and to fulfill the promise which was given by my Father before the foundation of the world; and for their exaltation in the eternal worlds, that they may bear the souls of men; for herein is the work of my father continued, that he may be glorified.

25. And again, verily, verily I say unto you, if any man have a wife who holds the keys of this power, and

he teaches unto her the law of my priesthood, as pertaining to these things; then shall she believe, and administer unto him, or she shall be destroyed, saith the Lord your God; for I will destroy her; for I will magnify my name upon all those who receive and abide in my law. Therefore, it shall be lawful in me, if she receive not this law, for him to receive all things whatsoever I the Lord his God, will give unto him, because she did not believe and administer unto him, according to my word; and she then becomes the transgressor, and he is exempt from the law of Sarah, who administered unto Abraham according to the law, when I commanded Abraham to take Hagar to wife. And now, as pertaining to this law: Verily, verily, I say unto you, I will reveal more unto you, hereafter; therefore, let this suffice for the present. Behold, I am Alpha and Omega. Amen.

CELESTIAL MARRIAGE.

A DISCOURSE DELIVERED BY ELDER ORSON PRATT, IN THE TABERNACLE, GREAT SALT LAKE CITY.

It is quite unexpected to me, brethren and sisters, to be called upon to address you this forenoon; and still more so, to address you upon the principle which has been named, namely, a plurality of wives.

It is rather new ground for me; that is, I have not been in the habit of publicly speaking upon this subject; and it is rather new ground to the inhabitants of the United States, and not only to them, but to a portion of the inhabitants of Europe; a portion of them have not been in the habit of preaching a doctrine of this description; consequently, we shall have to break up new ground.

It is well known, however, to the congregation before me, that the latter-day saints have embraced the doctrine of a plurality of wives, as a part of their religious faith. It is not, as many have supposed, a doctrine embraced by them to gratify the carnal lusts and feelings of man; that is not the object of the doctrine.

We shall endeavor to set forth before this enlightened assembly, some of the causes why the Almighty has revealed such a doctrine, and why it is considered a part and portion of our religious faith. And I believe that they will not, under our present form of government (I mean the government of the United States), try us for treason for believing and practising our religious notions and ideas. I think, if I am not mistaken, that the Constitution gives the privilege to all the inhabitants of this country, of the free exercise of their religious notions, and the freedom of their faith, and the practice of it. Then, if it can be proven to a demonstration, that the latter-day saints have actually embraced, as a portion of their religion, the doctrine of a plurality of wives, it is constitutional. And should there ever be laws enacted by this government to restrict them from the free exercise of this part of their religion, such laws must be unconstitutional.

But, says the objector, we cannot see how this doctrine can be embraced as a matter of religion and faith; we can hardly conceive how it can be embraced only as a kind of domestic concern, something that pertains to domestic pleasures, in no way connected with religion. In reply, we will show you that it is incorporated as a part of our religion, and necessary for our exaltation to the fullness of the Lord's glory in the eternal world. Would you like to know the reasons? Before we get through, we will endeavor to tell you why we consider it an essential doctrine to glory and exaltation, to our fullness of happiness in the world to come.

We will first make a few preliminary remarks in regard to the existence of man, to his first existence in his first estate; and then say something in relation to his present state, and the bearing which it has upon his next or future state.

The "Mormons" have a peculiar doctrine in regard to our pre-existence, different from the views of the

Christian world, so called, who do not believe that man had a pre-existence. It is believed, by the religious world, that man, both body and spirit, begins to live about the time that he is born into this world, or a little before; that then is the beginning of life. They believe that the Lord, by a direct act of creation, formed, in the first place, man out of the dust of the ground; and they believe that man is possessed of both body and spirit, by the union of which he became a living creature. Suppose we admit this doctrine concerning the formation of the body from the dust, then how was the spirit formed? Why, says one, we suppose it was made by a direct act of creation, by the Almighty Himself; that He moulded the spirit of man, formed and finished it in a proper likeness to inhabit the tabernacle He had made out of the dust.

Have you any account of this in the Bible? Do the Scriptures declare that the spirit was formed at the time the tabernacle was made? No. All the tabernacles of the children of men that were ever formed, from remote generations, from the days of Adam to this time, have been formed out of the earth. We are of the earth, earthy. The tabernacle has been organized according to certain principles and laws of organization, with bones, and flesh, and sinews, and skin. Now, where do you suppose all these tabernacles got their spirits? Does the Lord make a new spirit every time a tabernacle is made? If so, the work of creation, according to the belief of Christendom, did not cease on the seventh day. If we admit their views, the Lord must be continually making spirits to inhabit all the tabernacles of the children of men; He must make something like one thousand millions of spirits every century; He must be working at it every day, for there are many hundreds of individuals being born into the world every day. Does the Lord create a new spirit every time a new tabernacle comes into the world? That does not look reasonable, nor God-like.

But how is it, you inquire? Why the fact is, that being that animates this body, that gives life and energy, and power to move, to act, and to think: that being that dwells within this tabernacle is much older than what the tabernacle is. That spirit that now dwells within each man, and each woman, of this vast assembly of people, is more than a thousand years old, and I would venture to say, that it is more than five thousand years old.

But how was it made? when was it made? and by whom was it made? If our spirits existed thousands of years ago—if they began to exist—if there were a beginning to their organization, by what process was this organization carried on? Through what medium, and by what system of laws? Was it by a direct creation of the Almighty? Or were we framed according to a certain system of laws, in the same manner as our tabernacles? If we were to reason from analogy—if we admit analogical reasoning in the question, what would we say? We should say, that our spirits were formed by generation, the same as the body or tabernacle of flesh and bones. But what says revelation upon the subject? We will see whether revelation and analogy will agree.

We read of a certain time when the corner stones of the earth were laid, and the foundations thereof were made sure—of a certain time when the Lord began to erect this beautiful and glorious habitation, the earth then they had a time of joy. I do not know whether they had instruments of music, or whether they were engaged in the dance; but one thing is certain, they had great joy, and the heavens resounded with their shouts; yea, the Lord told Job, that all the sons of God shouted for joy, and the morning stars sang together, when the foundations of this globe were laid.

The SONS of God, recollect, shouted for joy, because there was a beautiful habitation being built, so that they could get tabernacles, and dwell thereon; they expected the time—they looked forward to the period; and it was joyful to them to reflect, that the creation was about being formed, the corner stone of it was laid, on which they might, in their times, and in their seasons, and in their generations, go forth and receive tabernacles for their spirits to dwell in. Do you bring it home to yourselves, brethren and sisters? Do you realize that you and I were there? Can you bring it to your minds that you and I were among that happy number that shouted for joy when this creation was made? Says one, I don't recollect it. No wonder! for your recollection is taken from you, because you are in a tabernacle that is earthly; and all this is right and necessary. The same is written of Jesus Christ himself, who had to descend below all things. Though he had wisdom to assist in the organization of this world; though it was through him, as the great leader of all these sons of God, the earth was framed, and framed too, by the assistance of all his younger brethren—yet we find, with all that great and mighty power he possessed, and the great and superior wisdom that was in his bosom, that after all his judgment had to be taken away; in his humiliation, his reason, his intelligence, his knowledge, and the power that he was formerly in possession of, vanished from him as he entered into the infant tabernacle. He was obliged to begin down at the lowest principles of knowledge, and ascend upward by degrees, receiving grace for grace, truth for truth, knowledge for knowledge, until he was filled with all the fullness of the Father, and was capable of ruling, governing, and controlling all things, having ascended above all things. Just so with us; we that once lifted up our united voices as sons and daughters of God, and shouted for joy at the laying of the foundation of this earth, have come here and taken tabernacles, after the pattern of our elder brother: and in our humiliation— for it is humiliation to be deprived of knowledge we once had, and the power we once enjoyed—in our humiliation, just like our elder brother, our judgment is taken away. Do we not read also in the Bible, that God is the Father of our spirits?

We have ascertained that we have had a previous existence. We find that Solomon, that wise man, says that when the body returns to the dust, the spirit returns to God who gave it. Now all of this congregation very well know, that if we never existed *there*, we could not *return* there. I could not return to California. Why? Because I never have been there. If you never were with the Father, the same as Jesus was before the foundation of the world, you never could return there, any more than I could to the West Indies, where I have never been. But if we have once been there, then we can see the force of the saying of the wise man, that the spirit returns to God who gave it —it goes back where it once was.

Much more evidence might be derived in relation to this subject, even from the English translation of the

Bible; but I do not feel disposed to dwell too long upon any particular testimony; suffice it to say, that the Prophet Joseph Smith's translation of the fore part of the book of Genesis is in print, and is exceedingly plain upon this matter. In this inspired translation we find the pre-existence of man clearly laid down, and that the spirits of all men, male and female, did have an existence, before man was formed out of the dust of the ground. But who was their Father? I have already quoted a saying that God is the Father of our spirits.

In one sense of the word, there are more Gods than one; and in another sense there is but one God. The scriptures speak of more Gods than one. Moses was called a God to Aaron, in plain terms; and our Saviour, when speaking upon this subject, says, "If the Scriptures called them Gods unto whom the word of God came, why is it that you should seek to persecute me, and kill me, because I testify that I am the son of God?" This in substance was the word of our Saviour; those to whom the word of God came, are called Gods, according to his testimony. All these beings of course are one, the same as the Father and the Son are one. The Son is called God, and so is the Father, and in some places the Holy Ghost is called God. They are one in power, in wisdom, in knowledge, and in the inheritance of celestial glory; they are one in their works; they possess all things, and all things are subject to them; they act in unison; and if one has power to become the Father of spirits, so has another; if one God can propagate his species, and raise up spirits after his own image and likeness, and call them his sons and daughters, so can all other Gods that become like him, do the same thing; consequently, there will be many Fathers, and there will be many families, and many sons and daughters; and they will be the children of those glorified celestial beings that are counted worthy to be Gods.

Here let me bring for the satisfaction of the saints, the testimony of the vision given to our Prophet and Revelator Joseph Smith, and Sidney Rigdon, on the 25th day of February, 1832. They were engaged in translating the New Testament, by inspiration; and while engaged in this great work, they came to the 29th verse of the 5th chapter of John, which was given to them in these words—"they who have done good, in the resurrection of the just; and they who have done evil in the resurrection of the unjust." This being given in different words from the English translation, caused them to marvel and wonder; and they lifted up their hearts in prayer to God, that He would show them why it was that this should be given to them in a different manner; and behold, the visions of heaven opened before them. They gazed upon the eternal worlds, and saw things before this world was made. They saw the spiritual creation who were to come forth and take upon themselves bodies; and they saw things as they are to be in the future; they saw the celestial, terrestrial, and telestial worlds, as well as the sufferings of the ungodly; all passed before him in this great and glorious vision. And while they were yet gazing upon things as they were before the world was made, they were commanded to write, saying, "this is the testimony, last of all, which we give of him, that he lives; for we saw him, even on the right hand of God; and we heard the voice bearing record that he is the Only Begotten of the Father; that by him, and through him, and of him, the worlds are and were created; and the inhabitants thereof are begotten sons and daughters unto God." Notice this last expression, "the inhabitants thereof are begotten sons and daughters unto God," (meaning the different worlds that have been created and made). Notice, this does not say, that God, whom we serve and worship, was actually the Father Himself, in His own person, of all these sons and daughters of the different worlds; but they " are begotten sons and daughters unto God;" that is, begotten by those who are made like Him, after His image, and in His likeness; they begat sons and daughters, and begat them *unto* God, to inhabit these different worlds we have been speaking of. But more of this, if we have time, before we get through.

We now come to the second division of our subject, or the entrance of these spirits upon their second estate, or their birth and existence in mortal tabernacles. We are told that among this great family of spirits, some were more noble and great than others, having more intelligence.

Where do you read that? says one. Out of the Book of Abraham, translated from the Egyptian papyrus by the Prophet Joseph Smith. Among the great and numerous family of spirits—" the begotten sons and daughters of God"—there are some more intelligent than others; and the Lord showed unto Abraham " the intelligences that were organized before the world was; and among all these there were many of the noble and great ones." And God said to Abraham, "thou art one of them, thou wast chosen before thou wast born." Abraham was chosen before he was born. Here then, is knowledge, if we had time to notice it, upon the doctrine of election. However, I may just remark, it does not mean unconditional election to eternal life of a certain class, and the rest doomed to eternal damnation. Suffice it to say, that Abraham and many others of the great and noble ones in the family of spirits, were chosen before they were born, for certain purposes, to bring about certain works, to have the privilege of coming upon the stage of action, among the host of men, in favorable circumstances. Some came through good and holy parentages, to fulfill certain things the Lord decreed should come to pass, from before the foundations of the world.

The Lord has ordained that these spirits should come here and take tabernacles by a certain law, through a certain channel; and that law is the law of marriage. There are a great many things that I will pass by; I perceive that if I were to touch upon all these principles, the time allotted for this discourse would be too short, therefore I am under the necessity of passing by many things in relation to these spirits in their first estate, and the laws that governed them there, and come to their second estate.

The Lord ordained marriage between male and female as a law through which spirits should come here and take tabernacles, and enter into the second state of existence. The Lord Himself solemnized the first marriage pertaining to this globe, and pertaining to flesh and bones here upon this earth. I do not say pertaining to mortality; for when the first marriage was celebrated, no mortality was there. The first marriage that we have any account of, was between two immortal beings —old father Adam and old mother Eve; they were immortal beings; death had no dominion, no power over them; they were capable of enduring for ever and ever, in their organization. Had they fulfilled the law, and kept within certain conditions and bounds, their tabernacles would never have been seized by death; death entered entirely by sin, and sin alone. This marriage was celebrated between two immortal beings. For how long? Until death? No. That was entirely out of the question; there could have been no such thing in the ceremony.

What would you consider, my hearers, if a marriage was to be celebrated between two beings not subject to death? Would you consider them joined together for a certain number of years, and that then all their covenants were to cease for ever, and the marriage contract be dissolved? Would it look reasonable and consistent? No. Every heart would say that the work of God is perfect in and of itself, and inasmuch as sin had not brought imperfection upon the globe, what God joined together could not be dissolved, and destroyed, and torn asunder by any power beneath the celestial world, consequently it was eternal; the ordinance of union was eternal; the sealing of the great Jehovah upon Adam and Eve was eternal in its nature, and was never instituted for the purpose of being overthrown and brought to an end. It is known that the "Mormons" are a peculiar people about marriage; we believe in marrying, not only for a time, but for all eternity. This is a curious idea, says one, to be married for all eternity. It is not curious at all; for when we come to examine the Scriptures, we find that the very first example set for the whole human family, as a pattern instituted for us to follow, was not instituted until death, for death had no dominion at that time; but it was an eternal blessing pronounced upon our first parents. I have not time to explain further the marriage of Adam and Eve, but will pass on to their posterity.

It is true, that they became fallen, but there is a redemption. But some may consider that the redemption only redeemed us in part, that is, merely from some of the effects of the fall. But this is not the case; every man and woman must see at once that a redemption must include a complete restoration of all the privileges lost by the fall.

Suppose, then, that the fall was of such a nature as to dissolve the marriage covenant, by death—which is not necessary to admit, for the covenant was sealed previous to the fall, and we have no account that it was dissolved —but suppose this was the case, would not the redemption be equally broad as the fall, to restore the posterity of Adam back to that which they lost? And if Adam and Eve were married for all eternity, the ceremony was an everlasting ordinance, that they twain should be one flesh for ever. If you and I should ever be accounted worthy to be restored back from our fallen and degraded condition to the privileges enjoyed before the fall, should we not have an everlasting marriage seal, as it was with our first progenitors? If we had no other reasons in all the Bible, this would be sufficient to settle the case at once in the mind of every reflecting man and woman, that inasmuch as the fall of man has taken away any privileges in regard to the union of male and female, these privileges must be restored in the redemption of man, or else it is not complete.

What is the object of this union? is the next question. We are told the object of it: it is clearly expressed; for, says the Lord unto the male and female, I command you to multiply and replenish the earth. And, inasmuch as we have proved that the marriage ordinance was eternal in its nature, previous to the fall, if we are restored back to what was lost by the fall, we are restored for the purpose of carrying out the commandment given before the fall, namely, to multiply and replenish the earth. Does it say, continue to multiply for a few years, and then the marriage contract must cease, and there shall be no further opportunity of carrying out this command, but it shall have an end? No, there is nothing specified of this kind; but the fall has brought in disunion through death; it is not a part of the original plan; consequently, when male and female are restored from the fall, by virtue of the everlasting and eternal covenant of marriage, they will continue to increase and multiply to all ages of eternity, to raise up beings after their own order, and in their own likeness and image, germs of intelligence, that are destined, in their times and seasons, to become not only sons of God, but gods themselves.

This accounts for the many worlds we heard Elder Grant speaking about yesterday afternoon. The peopling of worlds, or an endless increase, even of one family, would require an endless increase of worlds; and if one family were to be united in the eternal covenant of marriage, to fulfill that great commandment, to multiply his species, and propagate them, and if there be no end to the increase of his posterity, it would call for an endless increase of new worlds. And if one family calls for this, what would innumerable millions of families call for? They would call for as many worlds as have already been discovered by the telescope; yea, the number must be multiplied to infinity in order that there may be room for the inheritance of the sons and daughters of the Gods.

Do you begin to understand how these worlds get their inhabitants? Have you learned that the sons and daughters of God before me this day, are His offspring—made after His own image; that they are to multiply their species until they become innumerable.

Let us say a few words, before we leave this part of the subject, on the promises made to Abraham, Isaac, and Jacob. The promises were, Lift up your eyes, and behold the stars; so thy seed shall be, as numberless as the stars. What else did He promise? Go to the seashore, and look at the ocean of sand, and behold the smallness of the particles thereof, and then realize that your seed shall be as numberless as the sands. Now let us take this into consideration. How large a bulk of sand would it take to make as many inhabitants as there are now upon the earth? In about one cubic foot of sand, reckoning the grains of a certain size, there would be a thousand million particles. Now that is about the estimated population of our globe. If our earth were to continue 8,000 years, or eighty centuries, with an average population of one thousand millions per century, then three cubic yards of sand would contain a greater number of particles than the whole population of the globe, from the beginning, until the measure of the inhabitants of this creation is complete. If men then cease to multiply, where is the promise made to Abraham? Is it fulfilled? No. If that is the end of his increase, behold, the Lord's promise is not fulfilled. For the amount of sand representing his seed, might all be drawn in a one-horse cart; and yet the Lord said to Abraham, thy seed shall be as numerous as the sand upon the sea-shore; that is, to carry out the idea in full, it was to be endless; and therefore, there must be an infinity of worlds for their residence. We cannot comprehend infinity. But suffice it to say, if all the sands on the sea-shore were numbered, says the Prophet Enoch, and then all the particles of the earth besides, and then the particles of millions of earths like this, it would not be a beginning to all thy creations; and yet thou art there, and thy bosom is there; and thy curtains are stretched out still. This gives plenty of room for the fulfillment of the promise made to Abraham, and enough to spare for the fulfillment of similar promises to all his seed.

We read that those who do the works of Abraham, are to be blessed with the blessing of Abraham. Have you not, in the ordinances of this last dispensation, had the blessings of Abraham pronounced upon your heads? O yes, you say, I well recollect, since God has restored the everlasting Priesthood, that by a certain ordinance these blessings were placed upon our heads—the blessings of Abraham, Isaac, and Jacob. Why, says one, I never thought of it in this light before. Why did you not think of it? Why not look upon Abraham's blessings as your own, for the Lord blessed him with a promise of seed as numerous as the sand upon the sea-shore; so will you be blessed, or else you will not inherit the blessings of Abraham.

How did Abraham manage to get a foundation laid for this mighty kingdom? Was he to accomplish it all through one wife? No. Sarah gave a certain woman to him whose name was Hagar, and by her a seed was to be raised up unto him. Is this all? No. We read of his wife Keturah, and also of a plurality of wives and concubines, which he had, from whom he raised up many sons. Here, then, was a foundation laid for the fulfillment of the great and grand promise concerning the multiplicity of his seed. It would have been rather a slow process, if Abraham had been confined to one wife, like some of those narrow, contracted nations of modern Christianity.

I think there is only about one-fifth of the population of the globe, that believe in the one-wife system; the other four-fifths believe in the doctrine of a plurality of wives. They have had it handed down from time immemorial, and are not half so narrow and contracted in their minds as some of the nations of Europe and America, who have done away with the promises, and deprived themselves of the blessings of Abraham, Isaac, and Jacob. The nations do not know anything about the blessings of Abraham; and even those who have only one wife, cannot get rid of their covetousness, and get their little hearts large enough to share their property with a numerous family; they are so penurious, and so narrow and contracted in their feelings, that they take every possible care not to have their families large; they do not know what is in the future, nor what blessings they are depriving themselves of, because of the traditions of their fathers; they do not know that a man's posterity, in the eternal worlds, are to constitute his glory, his kingdom, and dominion.

Here, then, we perceive, just from this one principle, reasoning from the blessings of Abraham alone, the necessity—if we would partake of the blessings of Abraham, Isaac, and Jacob—of doing their works; and he that will not do the works of Abraham, and walk in his footsteps, will be deprived of his blessings.

Again, let us look at Sarah's peculiar position in regard to Abraham. She understood the whole matter; she knew that, unless seed was raised up to Abraham, he would come short of his glory; and she understood the promise of the Lord, and longed for Abraham to have seed. And when she saw that she was old, and fearing that she should not have the privilege of raising up seed, she gave to Abraham, Hagar. Would Gentile Christendom do such things now-a-days? O no; they would consider it enough to send a man to an endless hell of fire and brimstone. Why? Because tradition has instilled this in their minds as a dreadful, awful thing.

It matters not to them how corrupt they are in female prostitution, if they are lawfully married to only one wife; but it would be considered an awful thing by them to raise up a posterity from more than one wife; this would be wrong indeed; but to go into a brothel, and there debauch themselves in the lowest haunts of degradation all the days of their lives, they consider only a trifling thing; nay, they can even license such institutions in Christian nations, and it all passes off very well.

That is tradition; and their posterity have been fostered and brought up in the footsteps of wickedness. This is death, as it stalks abroad among the great and popular cities of Europe and America.

Do you find such haunts of prostitution, degradation, and misery here, in the cities of the mountains? No. Were such things in our midst, we should feel indignant enough to see that such persons be blotted out of the page of existence. These would be the feelings of this community.

Look upon those who committed such iniquity in Israel, in ancient days; every man and woman who committed adultery were put to death. I do not say that this people are going to do this; but I will tell you what we believe—we believe it ought to be done.

Whoredom, adultery, and fornication, have cursed the nations of the earth for many generations, and are increasing fearfully upon the community; but they must be entirely done away from those who call themselves the people of God; if they are not, woe! woe! be unto them, also; for "thus saith the Lord God Almighty," in the Book of Mormons, "Woe unto them that commit whoredoms, for they shall be thrust down to hell!" There is no getting away from it. Such things will not be allowed in this community; and such characters will find that the time will come, that God, whose eyes are upon all the children of men, and who discerneth the things that are done in secret, will bring their acts to light; and they will be made an example before the people; and shame and infamy will cleave to their posterity after them, unto the third and fourth generation of them that repent not.

How is this to be prevented? for we have got a fallen nature to grapple with. It is to be prevented in the way the Lord devised in ancient times; that is, by giving to His faithful servants a plurality of wives, by which a numerous and faithful posterity can be raised up, and taught in the principles of righteousness and truth; and then, after they fully understand those principles that were given to the ancient Patriarchs, if they keep not the law of God, but commit adultery, and transgressions of this kind, let their names be blotted out from under heaven, that they may have no place among the people of God.

But again, there is another reason why this plurality should exist among the latter-day saints. I have already given you one reason, and that is, that you might inherit the blessings and promises made to Abraham, Isaac, and Jacob, and receive a continuation of your posterity, that they may become as numerous as the sand upon the sea-shore. There is another reason, and a good one, too. What do you suppose it is? I will tell you; and it will appear reasonable to every man and woman of a reflecting mind. Do we not believe, as the Scriptures have told us, that the wicked nations of the earth are doomed to destruction? Yes, we believe it. Do we not also believe, as the Prophets have foretold, concerning the last days, as well as what the new revelations have said upon the subject, that darkness prevails upon the earth, and gross darkness upon the minds of the people; and not only this, but that all flesh has corrupted its way upon the face of the earth; that is, that all nations, speaking of them as nations, have corrupted themselves before the Most High God, by their wickedness, whoredoms, idolatries, abominations, adulteries, and all other kind of wickedness? And we furthermore believe, that according to the Jewish prophets, as well as the book of Mormon, and modern revelations given in the book of doctrine and covenants, that the sword of the vengeance of the Almighty is already unsheathed, and stretched out, and will no more be put back into the scabbard until it falls upon the head of the nations until they are destroyed, except they repent. What else do we believe? We believe that God is gathering out from among these nations those who will hearken to his voice, and receive the proclamation of the Gospel, to establish them as a people alone by themselves, where they can be instructed in the right way, and brought to a knowledge of the truth. Very well: if this be the case, that the righteous are gathering out, and are still being gathered from among the nations, and being planted by themselves, one thing is certain—that that people are better calculated to bring up their children in the right way, than any other under the whole heavens. Oh yes, says one, if that is the case—if you are the people the ancient Prophets have spoken of, if you are the people that are guided by the Lord, if you are under the influence, power, and guidance of the Almighty, you must be the best people under heaven, to dictate the young mind: but what has that to do with the plurality of wives? I will tell you. I have already told you that the spirits of men and women, all had a previous existence, thousands of years ago, in the heavens, in the presence of God; and I have already told you that among them are many spirits that are more noble, more intelligent than others, that were called the great and mighty ones, reserved until the dispensation of the fullness of times, to come forth upon the face of the earth, through a noble parentage that shall train their young and tender minds in the truths of eternity, that they may grow up in the Lord, and be strong in the power of His might, be clothed upon with His glory, be filled with exceeding great faith; that the visions of eternity may be opened to their minds; that they may be prophets, priests, and kings to the Most High God. Do you believe, says one, that they are reserved until the last dispensation, for such a noble purpose? Yes; and among the Saints is the most likely place for these spirits to take their tabernacles, through a just and righteous parentage. They are to be sent to that people that are the most righteous of any other people upon the earth; there to be trained up properly, according to their nobility and intelligence, and according to the laws which the Lord ordained before they were born. This is the reason why the Lord is sending them here, brethren and sisters; they are appointed to come and take their bodies here, that in their generations they may be raised up among the righteous. The Lord has not kept them in store for five or six thousand years past, and kept them waiting for their bodies all this time, to send them among the Hottentots, the African negroes, the idolatrous Hindoos, or any other of the fallen nations that dwell upon the face of this earth. They are not kept in reserve in order to come forth to receive such a degraded parentage upon the earth; no, the Lord is not such a being, His justice, goodness, and mercy will be magnified towards those who were chosen before they were born; and they long to come, and they will come, among the Saints of the living God; this would be their highest pleasure and joy, to know that they could have the privilege of being born of such noble parentage.

Then is it not reasonable, and consistent that the Lord should say unto His faithful and chosen servants, that had proved themselves before Him all the day long, that had been ready and willing to do whatsoever His will required them to perform;—take unto yourselves more wives, like unto the patriarchs, Abraham, Isaac, and Jacob, of old—like those who lived in ancient times, who walked in my footsteps, and kept my commands? Why should they not do this? Suppose the Lord should answer this question, would He not say, I have here in reserve, noble spirits, that have been waiting for thousands of years, to come forth in the fullness of times, and which I designed should come forth through these my faithful and chosen servants, for I know they will do my will, and they will teach their children after them to do it? Would not this be the substance of the language, if the Lord should give us an answer upon this subject?

But then another question will arise; how are these things to be conducted? Are they to be left at random? Is every servant of God at liberty to run here and there, seeking out the daughters of men as wives unto themselves without any restriction, law, or condition? No. We find these things were restricted in ancient times. Do you not recollect the circumstance of the Prophet Nathan's coming to David? He came to reprove him for certain disobedience, and told him about the wives he had lost through it; that the Lord would give them to another; and he told him, if he had been faithful, that the Lord would have given him still more, if he had only asked for them. Nathan the Prophet, in relation to David, was the man that held the keys concerning this matter in ancient days; and it was governed by the strictest laws.

So in these days; let me announce to this congregation, that there is but one man in all the world, at the same time, who can hold the keys of this matter; but one man has power to turn the key to inquire of the Lord, and to say whether I, or these my brethren, or any of the rest of this congregation, or the Saints upon the face of the whole earth, may have this blessing of

Abraham conferred upon them; he holds the keys of these matters now, the same as Nathan, in his day.

But, says one, how have you obtained this information? By new revelation. When was it given; and to whom? It was given to our Prophet, Seer, and Revelator, Joseph Smith, on the 12th day of July, 1843; only about eleven months before he was martyred for the testimony of Jesus.

He held the keys of these matters; he had the right to inquire of the Lord; and the Lord has set bounds and restrictions to these things; He has told us in that revelation, that only one man can hold these keys upon the earth at the same time; and they belong to that man who stands at the head to preside over all the affairs of the church and kingdom of God in the last days. They are the sealing keys of power, or in other words, of Elijah, having been committed and restored to the earth by Elijah, the prophet, who held many keys, among which were the keys of sealing, to bind the hearts of the fathers to the children, and the children to the fathers; together with all the other sealing keys and powers, pertaining to the last dispensation. They were committed by that angel who administered in the Kirtland Temple, and spoke unto Joseph the prophet, at the time of the endowments in that house.

Now, let us enquire, what will become of those individuals who have this law taught unto them in plainness, if they reject it? [A voice in the stand, "they will be damned."] I will tell you: they will be damned, saith the Lord God Almighty, in the revelation he has given. Why? Because where much is given, much is required; where there is great knowledge unfolded for the exaltation, glory, and happiness of the sons and daughters of God, if they close up their hearts, if they reject the testimony of His word, and will not give heed to the principles he has ordained for their good, they are worthy of damnation, and the Lord has said they shall be damned. This was the word of the Lord to His servant Joseph the prophet himself. With all the knowledge and light he had, he must comply with it, or, says the Lord unto him, you shall be damned; and the same is true in regard to all those who reject these things.

What else have we heard from our President? He has related to us that there are some damnations that are eternal in their nature; while others are but for a certain period, they will have an end, they will not receive a restoration to their former privileges, but a deliverance from certain punishments: and instead of being restored to all the privileges pertaining to man previous to the fall, they will only be permitted to enjoy a certain grade of happiness, not a full restoration. Let us inquire after those who are to be damned, admitting they will be redeemed, which they will be, unless they have sinned against the Holy Ghost. They will be redeemed, but what will it be to? Will it be to exaltation, and to a fullness of glory? Will it be to become sons of God, or Gods to reign upon thrones, and multiply their posterity, and reign over them as kings? No, it will not. They have lost that exalted privilege for ever; though they may, after having been punished for long periods, escape by the skin of their teeth; but no kingdom will be conferred upon them? What will be their condition? I will tell you what revelation says, not only concerning them that reject these things, but concerning those that through their carelessness, or want of faith, or something else, have failed to have their marriages sealed for time and for all eternity; those that do not do these things, so as to have the same ordinances sealed upon their heads by divine authority, as was upon the head of old Father Adam—if they fail to do it through wickedness, through their ungodliness, they also will never have the privilege of possessing that which is possessed by the Gods that hold the keys of power, of coming up to the thrones of their exaltation, and receiving their kingdoms. Why? Because, saith the Lord, all oaths, all covenants, and all agreements, etc., that have been made by man, and not by me, and by the authority I have established, shall cease when death shall separate the parties; that is the end; that is the cessation; they go no further; and such a person cannot come up in the morning of the resurrection, and say, Behold, I claim you as my wife; you are mine; I married you in the other world before death; therefore you are mine: he cannot say this. Why? Because he never married that person for eternity.

Suppose they should enter into covenant and agreement, and conclude between themselves to live together to all eternity, and never have it sealed by the Lord's sealing power, by the holy priesthood, would they have any claim on each other in the morning of the resurrection? No; it would not be valid nor legal, and the Lord would say, It was not by me; your covenants were not sealed on the earth, and therefore they are not sealed in the heavens; they are not recorded on my book; they are not to be found in the records that are in the archives of eternity; therefore, the blessings you might have had, are not for you to enjoy. What will be their condition? The Lord has told us. He says these are angels; because they keep not this law, they shall be ministering servants unto those who are worthy of obtaining a more exceeding and eternal weight of glory; wherefore, saith the Lord, they shall remain singly and separately in their saved condition, and shall not have power to enlarge themselves, and thus shall they remain forever and ever.

Here, then, you can read their history; they are not Gods, but they are angels or servants to the Gods. There is a difference between the two classes; the Gods are exalted; they hold keys of power; are made kings and priests; and this power is conferred upon them in time, by the everlasting priesthood, to hold a kingdom in eternity that shall never be taken from them worlds without end; and they will propagate their species. They are not servants; for one God is not to be a servant to another God; they are not angels; and this is the reason why Paul said, know ye not, brethren, that we shall judge angels? Angels are inferior to the saints who are exalted as kings. These angels who are to be judged, and to become servants to the Gods, did not keep the law, therefore, though they are saved, they are to be servants to those who are in a higher condition.

What does the Lord intend to do with this people? He intends to make them a kingdom of kings and priests, a kingdom unto Himself, or in other words, a kingdom of Gods, if they will hearken to His law. There will be many who will not hearken; there will be the foolish among the wise, who will not receive the new and everlasting covenant in its fullness; and they never will attain to their exaltation; they never will be counted worthy to hold the sceptre of power over a numerous progeny, that shall multiply themselves without end, like the sand upon the sea shore.

We can only touch here and there upon this great subject, we can only offer a few words with regard to this great, sublime, beautiful and glorious doctrine, which has been revealed by the Prophet, Seer, and Revelator, Joseph Smith, who sealed his testimony with his blood, and thus revealed to the nations, things that were in ancient times, as well as things that are to come.

But while I talk, the vision of my mind is opened; the subject spreads forth and branches out like the branches of a thrifty tree; and as for the glory of God, how great it is! I feel to say, Hallelujah to his great and holy name; for He reigns in the heavens, and He will exalt His people to sit with Him upon thrones **of** power, to reign forever and ever.

INDIAN HOSTILITIES AND TREACHERY—MORMON POLICY TOWARDS THEM.

AN ADDRESS DELIVERED BY PRESIDENT BRIGHAM YOUNG, IN THE TABERNACLE, GREAT SALT LAKE CITY.

I wish to say a few words to the latter-day saints this morning, as there seems to be considerable excitement in the feelings of the people, and many inquiring what will be the result of the present Indian difficulties.

I will give you my testimony, as far as I have one on the subject, concerning these difficulties in this territory, north and south, pertaining to our brethren, the Lamanites. My testimony to all is—IT IS RIGHT, and perfectly calculated, like all other providences of the Lord, of the like nature, to chasten this people until they are willing to take counsel. They will purify and sanctify the Saints, and prepare the wicked for their doom.

There has nothing strange and uncommon to man, yet occurred; nothing has yet happened out of the ordinary providences of the Lord. These common dealings of our great Head with His people have been manifested from days of old, in blessings and chastisements Wars, commotions, tumults, strife, nation contending against nation, and people against people, have all been governed and controlled by Him whose right it is to control such matters.

Among wicked nations, or among Saints, among the ancient Israelites, Philistines and Romans, the hand of the Lord was felt; in short, all the powers that have been upon the earth, have been dictated, governed, controlled, and the final issue of their existence has been brought to pass, according to the wisdom of the Almighty. Then my testimony is, IT IS ALL RIGHT.

There seems to be some excitement among the people, and fears are arising in the breasts of many, as to the general safety. Some person has been shot at by the Indians, or some Indians were seen in a hostile condition. And away go messengers to report to head quarters, saying, "*What shall we do?* for we cannot tell, but we shall all be killed by them; they have stolen our horses and driven off some cattle, which has created a great excitement in our settlement," etc; when, perhaps to-morrow, the very Indians who have committed these depredations will come and say, *How do you do?* We are friendly, cannot you give us some "*Chitcup?*"* They will shake hands, and appear as though it were impossible for them to be guilty of another hostility. And what is the next move? Why, our wise men, the elders of Israel, are either so fluctuating in their feelings, so unstable in their ways, or so ignorant of the Indian character, that the least mark of friendship manifested by these treacherous red men, will lull all their fears, throw them entirely off their guard, saying, "It is all right; wife, take care of the stock, for I am going to the cañon for a load of wood."

Away he goes, without a gun or a pistol to defend himself in case of an attack from some Indian or In-

---

* Indian language for something to eat.

dians, to rob him of his cattle and perhaps his life. Herds of cattle are driven upon the range, the feelings of the people are divested of all fear by this little show of Indian friendship, and their hearts are at peace with all mankind. They lie down to sleep at night with the doors of their houses open, and in many instances with no way to close them if they were willing, only by means of hanging up a blanket. Thus they go to sleep with their guns unloaded, and entirely without any means of defence, in case they should be attacked in the night. On the other hand, they no sooner discover an Indian in an hostile attitude, than the hue and cry is "*We shall all be murdered immediately.*" That is the kind of stability, the kind of unshaken self-command, the style of generalship and wisdom manifested by elders in Israel. To-day all are in arms, war is on hand; "we are going to be destroyed, or to fight our way through," is in every mouth. To-morrow all is peace, and every man turns to his own way, wherever the common avocations of life call him. No concern is felt as to protection in the future, but "all is right, all is safety, there is no fear of any further trouble," is the language of people's thoughts, and they lie down to sleep in a false security, to be murdered in the night by their enemies, if they are disposed to murder them.

I can tell you one thing with regard to excitement and war. You may take Israel here, as a community, with all their experience, and with all they have passed through in the shape of war, and difficulties of various kinds, and these wild Indians are actually wiser in their generation in the art of war than this people are. They lay better plans, display greater skill, and are steadier in their feelings. They are not so easily excited, and when excited, are not so easily allayed, as the men who have come, to inhabit these mountains, from where they have been trained and educated in the civilization of modern nations. You may not believe this assertion; it is, however, no matter whether you do or do not, the fact remains unaltered, as well as the conviction of my own mind regarding it.

I have been frequently asked, what is going to be the result of these troubles? I answer—*the result will be good*. What did you hear, you who have come to these valleys within the last few years, previous to your leaving your native country? You heard that all was peace and safety among the Saints in these regions; that the earth yielded in her strength, giving an abundance of food; and that this was a splendid country to raise stock. Your determination was then formed to go up to the valleys of the mountains, where you could enjoy peace and quiet, and follow the avocations of life undisturbed. When the people arrive here, many of them come to me and say, "Brother Brigham, can we go here or there to get us farms? Shall we enter into this or that speculation? We have been very poor, and we want to make some money, or we want the privilege of taking with us a few families to make a settlement in this or that distant valley." If I inquire why they cannot stay here, their answer is, "because there is no room, the land is chiefly taken up, and *we* have a considerable stock of cattle, we want to go where we can have plenty of range for our stock, where we can mount our horses, and ride over the prairies, and say, I am lord of all I survey. We do not wish to be disturbed, in any way, nor to be asked to pay tithing, to work upon the roads, nor pay territorial tax, but we wish all the time to ourselves, to appropriate to our own use. I want you, brother Brigham, to give us counsel that we can get the whole world in a string after us, and have it all in our own possession by and by." If there is light enough in Israel, let it shine in your consciences, and illuminate your understandings, and give you to know that I tell you the truth. This is the object many have, in wishing to settle and take in land that is far distant from the main body of the people. I have not given you the language of their lips to me, but the language of their hearts.

Elders of Israel are greedy after the things of this world. If you ask them if they are ready to build up the kingdom of God, their answer is prompt—" Why, to be sure we are, with our whole souls; but we want first to get so much gold, speculate and get rich, and then we can help the Church considerably. We will go to California and get gold, go and buy goods and get rich, trade with the emigrants, build a mill, make a farm, get a large herd of cattle, and *then* we can do a great deal for Israel." When will you be ready to do it? "In a few years, brother Brigham, if you do not disturb us. We do not believe in the necessity of doing military duty, in giving over our surplus property for tithing; we never could see into it; but we want to go and get rich, to accumulate and amass wealth, by securing all the land adjoining us, and all we have knowledge of." If that is not the spirit of this people, then I do not know what the truth is concerning the matter.

Now I wish to say to you who are fearing and trembling, do not be afraid at all, for it is certain if we should be killed off by the Indians, we could not die any younger; this is, about as good a time as can be for us to die, and if we all go together, why, you know, we shall have a good company along with us; it will not be lonesome passing through the valley, which is said to have a vail drawn over it. If we all go together, the dark valley of the shadow of death will be lighted up by us, so do not be scared. But there will not be enough slain by the Indians at this time to make the company very conspicuous in that dark valley. Do you begin to secretly wish you had stayed in the States or in England a little longer, until this Indian war had come to an end? There is a mighty fearing and trembling in the hearts of many. I know what men have done heretofore; when they have seen the enemy advancing, they have skulked, they were sure to be somewhere else than on hand when there was fighting to do, although, upon the whole, I have no fault to find with the latter-day saints, or with the Elders of Israel upon that subject, for they love to fight a little too well. If I were to have fears concerning them, it would not be that they would make war, but in the case of war being made on them, I should have more fear in consequence of the ignorant and foolish audacity of the elders, than of their being afraid. I should fear they would rush into danger like an unthinking horse into battle. So I will not find fault with regard to their courage. On that point I am a coward myself, and if people would do as I tell them, I would not only save my own life, but theirs likewise.

Suppose, now, that we should say to this congregation, and to all the wards of this city, *the time has come for us to fort up;* do you not think a great many persons would come immediately to me, and inquire if I did not think their houses quite safe enough, without being put to that trouble and expense? Yes, my office would be crowded with such persons, wanting to know if they might not live where they were now living, " for," they would say, " we have got good houses, and well finished off, besides, such a course will ruin them, and our gardens will go to destruction; we really cannot fort up." Would there not be a great amount of hard feelings upon the subject? I think so, whether you do or not. I think I should want as many as a legion of angels to assist me to convince every family it was necessary, if it actually was so.

I do not know but that the time may come, and that speedily, when I shall build a fort myself in this city, and those who are disposed can go into it with me, while the rest can stay out. When I see it is absolutely necessary to do this, I shall do it. If the people of Utah Territory would do as they were told, they would always be safe. If the people in San Pete county had done as they were told, from the beginning of that settlement, they would have been safe at this time, and would not have lost their cattle. The day before yesterday, Friday, July 29th, the Indians came from the mountains, to Father Allred's settlement, and drove off all the stock, amounting to two hundred head. If the people had done as they were told, they would not have suffered this severe loss, which is a just chastisement.

I recollect when we were down at Father Allred's settlement last April, they had previously been to me not only to know if they might settle in San Pete, but if they might separate widely from each other, over a piece of land about two miles square, each having a five acre lot for their garden, near their farms. They were told to build a good substantial fort, until the settlement became sufficiently strong, and not live so far apart, and expose themselves and their property to danger. Father

Allred told me they were then so nigh together, they did not know how to live! I told him they had better make up their minds to be baptized into the Church again, and get the Spirit of God, that each one might be able to live in peace with his neighbor in close quarters, and not think himself infringed upon. They wanted to know if they were to build a fort. "Why, yes," I said, "build a strong fort, and a corral to put your cattle, that the Indians cannot get them away from you." "Do you think, brother Brigham, the Indians will trouble us here?" they inquired. I said, "It is none of your business whether they will or not, but you will see the time that you will need such preparations." But I did not think it would come so quickly. There will more come upon this people to destroy them than they at present think of, unless they are prepared to defend themselves, which I shall not take time, this morning, to dwell upon. I said also to the brethren at Utah, "Do you make a fort, and let it be strong enough, that Indians cannot break into it." They commenced, and did not make even the shadow of a fort, for in some places there was nothing more than a line to mark where the approaching shadow would be. They began to settle round upon the various creeks and streamlets, and the part of a fort that existed was finally pulled up, and carried away somewhere else. I have told you, from the beginning, you would need forts, where to build them, and how strong. I told you six years ago, to build a fort that the *Devil could not get into*, unless you were disposed to let him in, and that would keep out the Indians. Excuse me for saying devil; I do not often use the old gentleman's name in vain, and if I do it, it is always in the pulpit, where I do all my swearing.

I make this apology because it is considered a sin to say devil, and it grates on refined ears.

I told the settlement in San Pete, at the first, to build a fort. They did not do it, but huddled together beside a stone quarry, without a place of common shelter where they could defend themselves, in case of an Indian difficulty. They had faith they could keep the Indians off. Well, now is the time to call it into exercise. They did, after a while, build a temporary fort at San Pete, which now shields them in a time of trouble.

When the brethren went to Salt Creek, they wanted to make a settlement there, and inquired of me if they might do so. I told them, no, unless they first built an efficient fort. I forbade them taking their women and children there, until that preparatory work was fully accomplished. Has it ever been done? No, but families went there and lived in wagons and brush houses, perfectly exposed to be killed. If they have faith enough to keep the Indians off, it is all right.

From the time these distant valleys began to be settled, until now, there has scarcely been a day but what I have felt twenty-five ton weight, as it were, upon me, in exercising faith to keep this people from destroying themselves; but if any of them can exercise faith enough for themselves, and wish to excuse me, I will take my faith back.

The word has gone out now, to the different settlements, in the time of harvest, requiring them to build forts. Could it not have been done last winter, better than now? Yes. Do you not suppose people will now wish they had built forts when they were told? If they do not, it proves what they have been all the time, shall I say fools? If that is too harsh a term, I will say they have been foolish. It is better for me to labor in building a house or a fort, to get out fencing timber, and wood to consume through the winter, when I have nothing else to do, and not be under the necessity of leaving my grain on the ground to do those things. Harvest is no time to build forts, neither is it the time to do it when we should be plowing and sowing.

Now the harvest is upon us, I wish to say a few words concerning it. I desire you to tell your neighbors, and wish them to tell their neighbors, and thus let it go to the several counties around—now is the time for women and children to assist in the harvest fields, the same as they do in other countries. I never asked this of them before; I do not now ask it as a general thing, but those employed in the expedition south, in the work of defending their brethern from Indian depredations, who have heavy harvests on hand, rather than suffer the grain to waste, let the women get in the harvest, and put it where the Indians cannot steal it. And when you go into the harvest field, carry a good butcher knife in your belt, that if an Indian should come upon you, supposing you to be unarmed, you would be sure to kill him.

Tell your neighbors of this, and go to work, men, women, and children, and gather in your grain, and gather it clean, leave none to waste, and put it where the Indians cannot destroy it.

Does this language intimate anything terrific to you? It need not. If you will do as you are told, you will be safe continually. Secure your bread stuff, your wheat, and your corn, when it is ripe, and let every particle of grain raised in these valleys be put where it will be safe, and as much as possible from vermin, and especially from the Indians, and then build forts.

INDIAN HOSTILITIES AND TREACHERY. 49

Let every man and woman who has a house make that house a fort, from which you can kill ten where you now only kill one, if the Indians come upon you. "Brother Brigham, do you really expect Indians to come upon us in this city?" This inquiry, I have no doubt, is at this moment in the hearts of a few, almost breathless with fear. Were I to answer such inquirers as I feel, I should say, it is none of your business; but I will say, you are so instructed, to see if you will do as you are told. Let your dwelling house be a perfect fort. From the day I lived where brother Joseph Smith lived, I have been fortified all the time so as to resist twenty men, if they should come to my house in the night, with an intent to molest my family, assault my person, or destroy my property; and I have always been in the habit of sleeping with one eye open, and if I cannot then sufficiently watch, I will get my wife to help me. Let an hostile band of Indians come round my house, and I am good for quite a number of them. If one hundred should come, I calculate that only fifty would be able to go to the next house, and if the Saints there used up the other fifty, the third house would be safe.

But instead of the people taking this course, almost every good rifle in the Territory has been traded away to the Indians, with quantities of powder and lead, though they waste it in various ways when they have got it. The whites would sell the title to their lives, for the sake of trading with the Indians.

They will learn better, I expect, by and by, for the people have never received such strict orders as they have got now. I will give you the pith of the last orders issued—" That man or family who will not do as they are told in the orders, are to be treated as strangers, yea, even as enemies, and not as friends." And if there should be such a contest, if we should be called upon to defend our lives, our liberty, and our possessions, we would cut such off the first, and walk over their dead bodies to conquer the foe outside.

Martial law is not enforced yet, although the whole Territory is in a state of war, apparently, but it is only the Utah [Indians] who have declared war on Utah [Territory]. Deseret has not yet declared war; how soon it will be declared is not for me to say; but we have a right, and it is our duty, to put ourselves in a state of self-defence.

The few families that settled in Cedar Valley, at the point of the mountains, were instructed to leave there, last spring. They have gone back again, upon their own responsibility, and now want to know what they must do. They have been told to do just as they have a mind to.

Those who have taken their wives and children in the cañons to live, have been told to remove them into the city; and if you want to make shingles, or do any other work that requires you to remain there, have your gun in a situation that an Indian cannot creep up and steal it from you before you are aware, that you can be good for a few Indians if they should chance to come upon you.

If I wished to live away from the body of the people, my first effort should be directed towards building a good and efficient fort. When new settlements were made in the eastern countries, they built them of timber, and they were called "block-houses." I would advise that every house in a new settlement should be made good for all the Indians that could approach it, with an intention to tear it down. If I did not do that, I would go to where I could be safe, I would take up my abode with the body of the people. I would take my family there at least. By taking this course, every person will be safe from the depredations of the Indians, which are generally committed upon the defenceless and unprotected portions of the community.

I know what the feelings of the generality of the people are, at this time—they think all the Indians in the mountains are coming to kill off the latter-day saints. I have no more fear of that, than I have of the sun ceasing to give light upon the earth. I have studied the Indian character sufficiently to know what the Indians are in war, I have been with them more or less from my youth upward, where they have often had wars among themselves. Let every man, woman, and child, that can handle a butcher knife, be good for one Indian, and you are safe.

I am aware that the people want to ask me a thousand and one questions, whether they have done it or not, touching the present Indian difficulties. I have tried to answer them all, in my own mind, by saying, *it will be just as the Lord will.*

How many times have I been asked in the past week, what I intend to do with Walker. I say, LET HIM ALONE, severely. I have not made war on the Indians, nor am I calculating to do it. My policy is to give them presents, and be kind to them. Instead of being Walker's enemy, I have sent him a great pile of tobacco to smoke when he is lonely in the mountains. He is now at war with the only friends he has upon this earth, and I want him to have some tobacco to smoke.

I calculate to pursue just such a course with the Indians, and when I am dictated to by existing circumstances, and the Spirit of the Lord, to change my course, I will do it, and not until then.

If you were to see Walker, do you think you would kill him? You that want to kill him, I will give you a mission to that effect. A great many appear very bold, and desire to go and bring me Walker's head, but they want all the people in Utah to go with them. I could point out thousands in this Territory who would follow these Indians, and continue to follow them, and leave the cattle to be driven off by the emigrants, and the grain to perish, and thus subject the whole community to the ravages of famine, and its consequent evils. I have been teased and teased by men who will come to me and say, "Just give me twenty-five, fifty, or a hundred men, and I will go and fetch you Walker's head." I do not want his head, but I wish him to do all the devil wants him to do, so far as the Lord will suffer him, and the devil to chastise this people for their good.

I say to the Indians, as I have often said to the mob, go your length. You say you are going to kill us all off, you say you are going to obliterate the latter-day saints, and wipe them from the earth, why don't you do it, you poor miserable curses? The mob only had power to drive the Saints to their duty, and to remember the Lord their God, and that is all the Indians can do. This people are worldly-minded, they want to get rich in earthly substance, and are apt to forget their God, the pit from which they were dug, and the rock from which they were hewn, every man turning to his own way. Seemingly the Lord is chastening us until we turn and do His will. What are you willing to do? Would you be willing to build a fort, and all go in there to live? I tell you, you would have a hell of your own, and devils enough to carry it on. Do you suppose you will ever see the time you would do that, and live at peace with each other, and have the spirit of the Lord enough to look each other in the face, and say, with a heart full of kindness, "Good morning, Mary," or "How do you do, Maria?" You will be whipped until you have the Spirit of the Lord Jesus Christ sufficiently to love your brethren and sisters freely, men, women, and children; until you can live at peace with yourselves, and with every family around you; until you can treat every child as though it were the tender offspring of your own body, every man as your brother, and every woman as your sister; and until the young persons treat the old with that respect due to parents, and all learn to shake hands, with a warm heart, and a friendly grip, and say, "God bless you!" from morning until evening; until each person can say, "I love you all, I have no evil in my heart to any individual, I can send my children to school with yours, and can correct your children when they do wrong, as though they were my own, and I am willing you should correct mine, and let us live together until we are a holy and sanctified society." There will always be Indians or somebody else to chastise you, until you come to that spot; so amen to the present Indian trouble, for it is all right. I am just as willing the rebellious of this people should be kicked, and cuffed, and mobbed, and hunted by the Indians, as not, for I have preached to them until I am tired. I will give nc more counsel to any person upon the duties of self-preservation; you can do as you please; if you will not preserve yourselves, I may reason with you until my tongue cleaves to the roof of my mouth, to no avail. Let the Lord extend the hand of benevolence to brother Walker, and he will make you do it by other means than exhortations given in mildness.

This very same Indian Walker has a mission upon him, and I do not blame him for what he is now doing; he is helping me do the will of the Lord to this people, he is doing with a chastening rod what I have failed to accomplish with soft words, while I have been handing out my substance, feeding the hungry, comforting the sick. But this has no effect upon this people, at all, my counsel has not been heeded, so the Lord is making brother Walker an instrument to help me, and perhaps the means that he will use will have their due effect.

Do you suppose I want to kill him? No. I should be killing the very means that will make this people do what we wanted them to do years ago.

There are hundreds of witnesses to bear testimony that I have counselled this people, from the beginning, what to do to save themselves, both temporally and spiritually.

In one of our orders issued lately, the southern settlements were advised to send their surplus cattle to this valley. No quicker had the news reached them, than our ears were greeted with one continued whine, which meant, "We are afraid *you* want them." So we did, to take care of them for you.

When Father Allred was advised to adopt measures to secure themselves and their property, he replied, "Oh, I do not think there is the least danger in the world; we are perfectly able to take care of our stock, and protect ourselves against the Indians." All right, I thought, let circumstances prove that.

Now, as difficulties surround them, they say to me, "Why, brother Brigham, if you had only told us what to do, we would have done it. Were we not always willing to take your counsel?" Yes, you are a great deal more willing to take it, than to obey it. If people are willing to carry out good counsel, they will secure themselves accordingly.

I have thought of setting a pattern, by securing myself; but were I to build a fort for myself and family, I should want about a legion of angels from the throne of God, to stay nine months with me, to get my folks willing to go into it. But I am so independent about it, I care not the snap of my finger for one of them. If my wives will not go into a place of security with me, it is all right, they can stay out, and I will go in and take my children with me. I say, I do not know but I may take a notion to set a pattern by building a fort; if I do, some one in in this city may follow my example, and then somebody else, etc., until we have a perfect city of forts.

"Brother Brigham, do you really think we shall ever need them?" YES, I DO. All the difficulties there is in the community this year, is not a drop in comparison to the heavy shower that will come. "Well, and where is it coming from?" From hell, where every other trouble comes from. "And who do you think will be the actors?" Why, the Devil and his imps. [W. W. Phelps in the stand. We could not do very well without a Devil.] No, sir, you are quite aware of that; you *know* we could not do without him. If there had been no devil to tempt Eve, she never would have got her eyes opened. We need a devil to stir up the wicked on the earth to purify the saints. Therefore, let the devils howl, let them rage, and thus exhibit themselves in the form of those poor, foolish Lamanites. Let them go on in their work, and do you not desire to kill them, until they ought to be killed, and then we will extinguish the Indian title, if it is required.

Did you never feel to pity them, on viewing their wretched condition? Walker, with a small band, has succeeded in making all the Indian bands in these mountains fear him. He has been in the habit of stealing from the Californians, and of making every train of emigrants that passed along the Spanish trail to California pay tithing to him. He finally began to steal children form those bands to sell to the Spaniards; and through fear of him, he has managed to bring in subjection almost all the Utah tribes.

I will relate one action of Walker's life, which will serve to illustrate his character. He, with his band, about last February, fell in with a small band of Payedes, and killed off the whole of the men, took the squaws prisoners, and sold the children to the Mexicans, and some few were disposed of in this Territory. This transaction was told by Arapeen, Walker's brother, though he was not at the affray himself.

The Indians in these mountains are continually on the decrease; bands that numbered 150 warriors when we first came here, number not more than 35 now; and some of the little tribes in the southern parts of this Territory, towards New Mexico, have not a single squaw amongst them, for they have traded them off for horses, etc. This practice will soon make the race extinct. Besides, Walker is continually, whenever an opportunity presents itself, killing and stealing children from the wandering bands that he has any power over, which also has its tendency to extinguish the race.

Walker is hemmed in, he dare not go into California again. Dare he go east to the Snakes? No. Dare he go north? No, for they would rejoice to kill him. Here he is, penned up in a small compass, surrounded by his enemies; and now the elders of Israel long to eat up, as it were, him and his little band. What are they? They are a set of cursed fools. Do you not rather pity them? They dare not move over a certain boundary, on any of the four points of the compass, for fear of being killed; then they are killing one another, and making war upon this people that could use them up, and they not be a breakfast spell for them if they felt so disposed. See their condition, and I ask you, do you not pity them? From all appearance, there will not be an Indian left, in a short time, to steal a horse. Are they not fools, under these circumstances, to make war with their best friends?

Do you want to run after them to kill them? I say, let them alone, for peradventure God may pour out His Spirit upon them, and show them the error of their ways. We may yet have to fight them, though they are of the house of Israel to whom the message of salvation is sent; for their wickedness is so great, that the Lord Almighty cannot get at the hearts of the older ones to teach them saving principles. Joseph Smith said we should have to fight them. He said, "When this people mingle among the Lamanites, if they do not bow down in obedience to the Gospel, they will hunt them until there is but a small remnant of them left upon this continent." They have either got to bow down to the Gos pel or be slain. Shall we slay them simply because they will not obey the Gospel? No. But they will come to us and try to kill us, and we shall be under the necessity of killing them to save our own lives.

I wished to lay these things before the people this morning, to answer a great many questions, and allay their fears. Yesterday, brother Kimball heard at his mill, ten miles north, that I had sent word to him, that the mountains were full of Indians, and he and the families with him were to move into the city; so they immediately obeyed this report. Brother Kimball came to me and inquired if I had sent such orders. I said, no. But it is all right, for I wanted the women and children from there. This shows the excited state of the people.

One thing more. I ask you men who have been with Joseph in the wars *he* passed through, and who were with him at the time of his death, what was it that preserved us, to all outward appearances? It is true, in reality, God did it. But by what means did He keep the mob from destroying us? *It was by means of being well armed with the weapons of death to send them to hell cross lots.* Just so you have got to do.

And for this people fostering to themselves that the day has come for them to sell their guns and ammunition to their enemies, and sit down to sleep in peace, they will find themselves deceived, and before they know, they will sleep until they are slain. They have got to carry weapons with them, to be ready to send their enemies to hell cross lots, whether they be Lamanites, or mobs who may come to take their lives, or destroy their property. We must be so prepared that they dare not come to us in a hostile manner without being assured they will meet a vigorous resistance, and ten to one they will meet their grave.

The Lord will suffer no more trouble to come upon us than is necessary to bring this people to their senses. You need not go to sleep under the impression that it is the north and south only that is in danger, and we are all safe here. Now *mind*, let this people here lie down to sleep, and be entirely off their watch, and the first thing they know, they are in the greatest danger. You must not desert the watch tower, but do as I do—keep some person awake in your house all night long, and be ready, at the least tap of the foot, to offer a stout resistance, if it is required. Be ready at any moment to kill twenty of your enemies at least. Let every house be a fort.

After the cattle were stolen at San Pete, a messenger arrived here in about thirty hours to report the affair, and obtain advice. I told brother Wells, "you can write to them, and say, 'Inasmuch as you have no cows and oxen to trouble you, you can go to harvesting, and take care of yourselves.'" If you do not take care of yourselves, brethren, you will not be taken care of. I take care of them that hel**p** themselves. I will help you that try to help yourselves, and carry out the maxim of Doctor Dick—"God helps them that help themselves."

I am my own policeman, and have slept, scores of nights, with my gun and sword by my side, that is, if I slept at all. I am still a policeman. Now is the day to watch. It is as important for me to watch now, as well as pray, as it ever has been since I came into this kingdom. It requires watching, as well as praying men; take turns at it, let some watch while others pray, and then change round, but never let any time pass without a watcher, lest you be overtaken in an hour when you think not; it will come as a thief in the night. Look out for your enemies, for we know not how they will come, and what enemy it will be. Take care of yourselves.

Again, let me reiterate to the sisters, do not be afraid of going into the harvest field. If you are found there helping your, sons, your husbands and your brethren, to gather in the harvest, I say, God bless you, and I will also.

Take care of your grain, and take care of yourselves, that no enemy come to slay you. Be always on hand to meet them with death, and send them to hell, if they come to you. May God bless you all. Amen.

USE AND ABUSE OF BLESSINGS.

AN ADDRESS DELIVERED BY PRESIDENT B. YOUNG, IN THE TABERNACLE, GREAT SALT LAKE CITY.

I FEEL disposed to say a few words on the present occasion. It is said, that " at the sight of the eyes the heart is made to rejoice." This is truly the case with me this afternoon, when I look upon the congregation to see this spacious hall filled with the Saints of the Most High, for the purpose of partaking of the Sacrament of the Lord's Supper. It is a sight which I have not had the privilege of seeing before, only on Conference days. This morning I looked round to see how the house was crowded, which was packed to that extent, that scores could not be seated. I looked if peradventure I could designate any person that did not belong to the Church, that did not profess to be a Saint; but I could not see a single person of that description, that I knew of. I thought, why not be as diligent to attend the afternoon meetings, to partake of the Sacrament of the Lord's Supper, as to attend the morning meetings? Hitherto it has not been the case, but my heart rejoices to see the house so well filled this afternoon. I feel in my heart to bless you: it is full of blessings and not cursings. It is something that does not occupy my feelings to curse any individual, but I will modify this by saying *those who ought not to be cursed.* Who ought to be? Those who know their master's will, and do it not; they are worthy of many stripes; it is not those who do not know, and do not do, but those who know it, and do not do it—they are the ones to be chastised.

While the brethren have been speaking upon the blessings the Lord bestows upon this people, my mind has reflected upon many of the circumstances of life, and upon certain principles. I will ask you a question—*Do you think persons can be blessed too much?* I will answer it myself. Yes, they can, they can be blessed to their injury. For instance, suppose a person should be blessed with the knowledge of the holy Gospel, whose heart is set in him to do evil. We esteem this as a blessing, and would not the Lord consider it a blessing to bestow His favors and mercies upon any individual, by giving him a knowledge of life and salvation? But suppose He bestowed it upon persons whose hearts were set in them to do evil, who would by their wickedness turn these blessings into curses, they would be blessed too much. It is possible to bless people to death, you can bless them to everlasting misery by heaping too many blessings upon them. Perhaps this is what is meant by the saying—It is like heaping coals of fire upon their heads; it will injure them, consume them, burn them, destroy them. Suffice it to say, that people can be blessed too much. Can you bless a wise man too much? a man who knows what to do with his blessings when they are bestowed upon him? No, you cannot. Can you bless a wise people too much? No, it is impossible, when they know how to improve upon all blessings that are bestowed upon them. But the Lord does and will bless the inhabitants of the earth with such great and inestimable blessings, in the proclamation of the Gospel, that they will be damned who reject them, for light brings condemnation to men who love darkness rather than light.

Have this people been blessed too much? I will not positively say, but I think they have, inasmuch as their blessings in some instances have been to their injury. Why? Because they have not known what to do with their blessings.

While the brethren were speaking of the liberal hand of Providence in bestowing abundantly the products of the earth, it occurred to me, that this people, to my certain knowledge, had felt that they had too much, and they esteemed it as good for nothing. It is true what brother Jedediah Grant said with regard to wheat, and other grains, for I have seen it myself. I have seen hundreds, and thousands, and scores of thousands of bushels of grain lying to waste and rot, when it has not brought a great price. Many of this people have thought, and expressed themselves in language like this —"I can go to California, and get so much gold, or I can trade and make so much gold, I cannot therefore spend time to take care of wheat, nor to raise it; let it lie there and rot while I go and accumulate riches," They were then wealthy, for their granaries and barns were full of the blessings of the Lord, *but now they are empty*, because they did not know what to do with their blessings.

I can tell this people how to dispose of all their blessings, if they will only allow me time enough; and if I cannot tell them how, I can show them. For instance, you who have fields of wheat, beyond the limits of grasshoppers. will have considerable crops when it is harvested and perhaps so much that you will not know wha to with it. I know what you ought to do with it; you ought to say to your poor brethren—" Come and help take care of my grain, and share with me, and feed yourselves and your families." If you have so much that you cannot take care of it, and your neighbor is not without bread, tell Bishop Hunter that you have got so many hundred bushels to lay over in the store, and you will have the benefit of it on your tithing. That is what I recommend you to do with your blessings, when you have more than you can take care of yourselves. I say, hand it over and let your neighbors take care of it for you.

This makes me think of what I saw the first year I came into this valley, the same year I moved my family, which was the next season after the pioneers arrived here. It was late in the season when I arrived, but from the ground where this house now stands, there had been two crops of wheat. They had harvested the first crop very early, and the water being flooded over, it again started from the roots, and produced a fair crop, say from ten to twelve bushels to the acre. That was harvested, and it was coming up again. I said to the brethren, "Let these my brethren who have come with me gather up this wheat," but they would not suffer *them* to do it. Some of the brethren had gathered their crops of grain, and left a great deal wasting on the fields, I said, "Let the poor brethren, who have come in from abroad, glean in your fields." You can bear me witness that a great many widows and poor men came here, and brought but very little with them, and there never was a man, to my knowledge, ever expressed a desire to let them glean in his field. "All right," I said, "we

USE AND ABUSE OF BLESSINGS. 65

can live on greens," while at the same time there was more wasted that season than to make up the deficiency, that all might have been comfortable. Late in the fall I saw one man working among his corn; he had a large crop, more than a single man could take care of. I saw he was going to let it go to waste; I said to him, "Brother, let the brethren and sisters help you to husk your corn, to gather it and put it safely away, for so much it will benefit them and help you. "O," he replied, "I have nothing to spare, I can take care of it myself." I saw it wasting, and said to him, "Brother, get your corn husked immediately, and let the brethren do it, and pay them with a portion of it." He replied, "I cannot spare a bit of it." I have no question of it at all in my mind, but three-fourths of his corn went into the mud, and was trampled down by the cattle; and women and children went without bread in consequence of it. That man had no judgment, he knew not what to do with his blessings the Lord had bestowed upon him.

Were I to ask the question, how much wheat or anything else a man must have to justify him in letting it go to waste, it would be hard to answer; figures are inadequate to give the amount. *Never let anything go to waste.* Be prudent, *save everything*, and what you get more than you can take care of yourselves, ask your neighbors to help you. There are scores and hundreds of men in this house, if the question were asked them if they considered their grain a burthen and a drudge to them, when they had plenty last year and the year before, that would answer in the affirmative, and were ready to part with it for next to nothing. How do they feel now, when their granaries are empty? If they had a few thousand bushels to spare now, would they not consider it a blessing? They would. Why? Because it would bring the gold and silver. But pause for a moment, and suppose you had millions of bushels to sell and could sell it for twenty dollars per bushel, or for a million of dollars per bushel, no matter what amount, so that you sell all your wheat, and transport it out of the country, and you are left with nothing more than a pile of gold, what good would it do you? You could not eat it, drink it, wear it, or carry it off where you could have something to eat. The time will come that gold will hold no comparison in value to a bushel of wheat. Gold is not to be compared with it in value. Why would it be precious to you now? Simply because you could get gold for it? Gold is good for nothing, only as men value it. It is no better than a piece of iron, a piece of limestone, or a piece of sandstone, and it is not half so good as the soil from which we raise our wheat, and other necessaries of life. The children of men love it, they lust after it, are greedy for it, and are ready to destroy themselves and those around them, over whom they have any influence, to gain it.

When this people are blessed so much that they consider their blessings a burden and a drudge to them, you may always calculate on a cricket war, a grasshopper war, a drought, too much rain, or something else to make the scales preponderate the other way. This people have been blessed too much, so that they have not known what to do with their blessings.

What do we hear from the inhabitants of the different settlements? The cry is—"I do not wish to live out yonder, for there is no chance to speculate and trade with the emigrants." Have you plenty to eat? Have you plenty of wheat, fowls, butter, cheese, and calves? Are you not raising stock in abundance for flesh meat of different kinds? What use is gold if you get enough to eat, drink, and wear without it? . What is the matter? "Why, we are away off, and cannot get rich all at once." You are lusting after that which you do not know what to do with, for few men know what to do with riches when they possess them. The inhabitants of this valley have proved it. They have proved it. by their reckless waste of the products of the earth, by their undervaluing the blessings conferred upon them by the emigration, which has administered clothing and other necessaries to them. We can see men who can clothe themselves and their families easily, go into the cañons in their broad cloth pantaloons to get wood, or you may see them take a horse, and ride bare-backed until they tare them to pieces, that they are not fit to come to meeting in. They do not know how to take care of good clothing. Again, if we were digging in a water-ditch to-morrow, that required all hands, in consequence of the rising of the water, I have no doubt but you would see what I saw the other day—one of our young dandies, who was perhaps not worth the shirt on his back, came to work in a water-ditch, dressed in his fine broad-cloth pantaloons, and a fine bosomed shirt, and I have no doubt he would have worn gloves too if he had been worth a pair. You would see men of this description, who are without understanding, whole hearted, good fellows, and ready to do anything for the advancement of the public good, commence to dig in the mud and wet, in their fine clothes, and go into the water, up to their knees, with their fine calf-skin boots. This is a wanton waste of the blessings of God, that cannot be justifiable in His eyes, and in the eyes of prudent, thinking men, under ordinary circumstances. If prudence and economy are necessary at one time more than another, it is when a family or a nation are thrown upon its own resourses, as we are. But you may trace the whole lives of some men, and it will be impossible for you to point out a single portion of time, when they knew how to appreciate and how to use even the common comforts of life, when they had them, to say nothing of an abundance of wealth.

Again, there have been more contention and trouble between neighbors, in these valleys, with regard to surplus property, which was not needed by this people, than any other thing. For instance, a widow woman comes in here from the United States, and turns out on the range beyond Jordan, three yoke of oxen and a few cows, for she considers she is too poor to have them herded. Again, a man comes in with ten yoke of oxen; he also turns them out to wander where they please. If he is asked why he does not put them in a herd, he will tell you, " I do not want to pay the herding fee." Another comes on with three or four span of horses, and twenty or thirty yoke of cattle. Has he any for sale? No, but he turns them all out upon the range and they are gone. By and by he sends a boy on horseback to hunt them, who is unsuccessful in finding them after a week's toil. The owner turns out himself, and all hands, to hunt up his stock, but they also fail in finding them, they are all lost except a very few. He was not able to have them herded, he thought, though he possessed so much property, and knew nothing more than to turn them out to run at large. Thus he consumes his time, running after his lost property. He frets his feelings, for his mind is continually upon it; he is in such a hurry in the morning to go out to hunt his stock, that he has no time to pray; when he returns home late at night, worn out with toil and anxiety of mind, he is unfit to pray; his cattle are lost, his mind is unhinged and darkened through the neglect of his duty, and apostacy stares him in the face, for he is not satisfied with himself, and murmurs against his brethren, and against his God. By and by some of his cattle turn up with a strange brand upon them; they have been taken up, and sold to this person or that one. This brings contention and dissatisfaction between neighbor and neighbor. Such a person has too much property, more than he knows what to do with. It would be much better for a man who is a mechanic, and intends to follow his business, to give one out of two cattle which he may possess, to some person, for taking care of the other. It would be better for those who possess a great quantity of stock, to sell half of them to fence in a piece of land, to secure the other half, than to drive them all out to run at large, and lose three-fourths of them. If there are half-a-dozen men round me, and I can put a cow in their way or anything else that will do them any good, for fencing up a lot for me, the property I thus pay is not out of the world, but is turned over to those men who had not a mouthful of meat, butter, or milk; it is doing them good, and I am reaping the profit and benefit of their labors in exchange. If I did not do this, I must either see them suffer, or make a free distribution of a part of what I have among them.

It is impossible for me to tell you how much a man must possess to entitle him to the liberty of wasting anything, or of letting it be stolen and run away with by the Indians. The surplus property of this community, as poor as we are, has done more real mischief than everything else besides.

I will propose a plan to stop the stealing of cattle in coming time, and it is this—let those who have cattle on hand join in a company, and fence in about fifty thousand acres of land, make a dividend of their cattle, and appropriate what they can spare, to fence in a large field, and this will give employment to immigrants who are coming in. When you have done this, then get up another company, and so keep on fencing until all the vacant land is substantially enclosed.

Some persons will perhaps say—"I do not know how good and how high a fence it will be necessary to build to keep thieves out. I do not know either, except you build one that will keep out the devil. Build a fence which the boys and the cattle cannot pull down, and I will ensure you will keep your stock. Let every man lay his plans so as to secure for his present necessities, and hand over the rest to the laboring man; keep making improvements, building, and making farms, and that will not only advance his own wealth, but the wealth of the community.

A man has no right with property, which, according to the laws of the land, legally belongs to him, if he does not want to use it; he ought to possess no more than he can put to usury, and cause to do good to himself and his fellow-man. When will a man accumulate money enough to justify him in salting it down, or, in other words, laying it away in the chest, to lock it up, there to lie, doing no manner of good either to himself or his neighbor? It is impossible for a man ever to do it. No man should keep money or property by him that he cannot put to usury for the advancement of that property in value or amount, and for the good of the community in which he lives; if he does, it becomes a dead weight upon him, it will rust, canker, and gnaw his soul, and finally work his destruction, for his heart is set upon it. Every man who has got cattle, money, or wealth of any description, bone and sinew, should put it out to usury. If a man has the arm, body, head, the component parts of a system to constitute him a laboring man, and has nothing in the world to depend upon but his hands, let him put them to usury. Never hide up anything in a napkin, but put it forth to bring an increase. If you have got property of any kind, that you do not know what to do with, lay it out in making a farm, or building a saw mill or a woollen factory, and go to with your mights to put all your property to usury.

If you have more oxen and other cattle than you need, put them in the hands of other men, and receive their labor in return, and put that labor where it will increase your property in value.

I hope you will now lay your plans to set men to work who will be in here by and by, for there will be a host of them, and they will all want employment, who trust to their labor for a subsistence; they will all want something to eat, and calculate to work for it. In the first place keep the ground in good order to produce you plentiful crops of grain, and vegetables, and then take care of them.

Let me say to the sisters, those who have children, never consider that you have bread enough around you to suffer your children to waste a crust or a crumb of it. If a man is worth millions of bushels of wheat and corn, he is not wealthy enough to suffer his servant girl to sweep a single kernel of it into the fire; let it be eaten by something, and pass again into the earth, and thus fulfill the purpose for which it grew. Some mothers would fill a basket full of bread to make a plaything for their children, but I have not had flour enough in the time of my greatest abundance, to let my children waste one morsel of bread with my consent. No, I would rather feed the greatest enemy I have on the earth with it, than have it go into the fire. *Remember it*, do not waste anything, but take care of everything; save your grain, and make your calculations, so that when the brethren come in from the United States, from England, and other places, you can give them some potatoes, onions, beets, carrots, parsnips, watermelons, or anything else which you have, to comfort them, and cheer up their hearts, and if you have wheat, dispose of it to them, and receive their labor in return. Raise enough and to spare of all the staple necessaries of life, and lay your plans to hire your brethren who will come in this fall to fence your farms, improve your gardens, and make your city lots beautiful. Lay your plans to secure enough to feed yourselves, and one or two of the brethren that are coming to dwell with us.

When we first came into the valley, the question was asked me, if men would ever be allowed to come into this Church, and remain in it and hoard up their property. *I say*, NO. That is a short answer, and it is a pointed one. The man who lays up his gold and silver who caches it away in a bank, or in his iron safe, or buries it up in the earth, and comes here, and professes to be a Saint, would tie up the hands of every individual in this kingdom, and make them his servants if he could. It is an unrighteous, unhallowed, unholy, covetous principle; it is of the devil, and is from beneath. Let every person who has capital, put it to usury. Is he required to bring his purse to me, to any of the twelve, or to any person whatever, and lay it at their feet? No, not by me. But I will tell you what to do with your means. If a man comes in the midst of this people with *money*, let him use it in making improvements, in building, in beautifying his inheritance in Zion, and in increasing his capital by thus putting out his money to usury. Let him go and make a great farm, and stock it well, and fortify it all around with a good and efficient fence. What for? Why for the purpose of spending his money. Then let him cut it up into fields, and adorn it with trees, and build a fine house upon it. What for? Why for the purpose of spending his money. What will he do when his money is gone? The money thus spent with a wise and prudent hand, is in a situation to accumulate and increase a hundred fold. When he has done making his farm, and his means still increase by his diligent use of it, he can then commence and build a woollen factory, for instance, he can send and buy the sheep and have them brought here, have them herded here, and shear them here, and take care of them, then set the boys and girls to cleaning, carding, spinning, and weaving the wool into cloth, and thus employ hundreds and thousands of the brethren and sisters who have come from the manufacturing districts of the old country, and have not been accustomed to dig in the earth for their livelihood, who have not learned anything else but to work in the factory. This would feed them, and clothe them, and put within their reach the comforts of life; it would also create at home a steady market for the produce of the agriculturist, and the labor of the mechanic. When he has spent his hundred and fifty thousand dollars, which he began business with, and fed five hundred persons, from five to ten years, besides realizing a handsome profit from the labor of the hands employed, by the increased population, and consequent increased demand for manufactured goods, at the end of ten years his factory would be worth five hundred thousand dollars. Suppose he had wrapped up his hundred and fifty thousand in a napkin, for fear of losing it, it would have sent him down to perdition, for the principle is from beneath. But when he puts forth his money to usury, not to me or any other person, but where it will redouble itself, by making farms, building factories for the manufacture of every kind of material necessary for home consumption, establishing blacksmiths' shops and other mechanical establishments, making extensive improvements to beautify the whole face of the earth, until it shall become like the garden of Eden, it becomes a saving blessing to him and those around him. And when the kings, princes, and rulers of the earth shall come to Zion, bringing their gold, and silver, and precious stones with them, they will admire and desire your possessions, your fine farms, beautiful vineyards, and splendid mansions. They will say—*"We have got plenty of money, but we are destitute of such possessions as these."* Their money loses its value in their eyes when compared with the comfortable possessions of the Saints, and they will want to purchase your property. The industrious capitalist inquires of one of them—"Do you want to purchase this property? I have obtained it by my economy and judgment, and by the labor of my brethren, and in exchange for their labor I have been feeding and clothing them, until they also have comfortable situations, and means to live. I have this farm, which I am willing to sell to enable me to advance my other improvements." "Well," says the rich man, "how much must I give you for it?" "Five hundred thousand dollars," and perhaps it has not cost him more than one hundred thousand. He takes the money and builds up three or four such farms, and employs hundreds of his brethren who are poor.

Money is not real capital, it bears the title only. True capital is labor, and is confined to the laboring classes. They only possess it. It is the bone, sinew, nerve, and muscle of man that subdue the earth, make it yield its strength and administer to his varied wants. This power tears down mountains and fills up valleys, builds cities and temples, and paves the streets. In short, what is there that yields shelter and comfort to civilized man, that is not produced by the strength of his arm the making elements bend to his will?

I will not ask the question again—*How much must a man possess to authorize him to waste anything?* Three or four years ago money was of little value in this country; you might go round exhibiting a back load of gold, and hold out a large piece to a man, I was going to say, almost as big as this Bible, and ask him to work for you, but he would laugh at your offer, and tell you he was looking for some one to work for him. He would then hail another man who had been in Nauvoo, and passed through the pinches there, and had scarcely a shirt to his back, but he would reply—"I was looking for some man to work for me." Gold could not purchase labor, it was no temptation whatever, but those times are passed. It is not now as it was then. I consequently alter my counsel to the brethren. I used to command you to hand over your surplus property, or that which you could not take care of, to me, and I would apply it to a good purpose, but now I counsel you to put it into the hands of men who have nothing at all, and let them pay you for it in labor.

I have never been troubled with thieves stealing my property. If I am not smart enough to take care of what the Lord lends me, I am smart enough to hold my tongue about it, until I come across the thief myself, and then I am ready to tie a string round his neck.

I have not the least hesitation in saying that the loose conduct, and calculations, and manner of doing business, which have characterized men who have had property in their hands, have laid the foundation to bring our boys into the spirit of stealing. You have caused them to do it, you have laid before them every inducement possible, to learn their hands, and train their minds to take that which is not their own. Those young men who have been taken up the past season and condemned to ignominious punishment, may trace the cause of their shame to that foundation. *Distribute your property.* The man that thinks he requires ten yoke of cattle, and can only use one yoke, is laboring under a mistake, he ought to let nine yoke go to the laboring community. If every man would do this with the property which he is not using, all would be employed and have sufficient. This would be the most effectual means of bringing the vile practice of stealing cattle and other property to a termination, which, as I have already said, has been encouraged by covetous, selfish men, who have refused to use their property for their own good, or the community's.

Let us hold before our mind the miser. If the people of this community feel as though they wanted the whole world to themselves, hate any other person to possess anything, and would hoard up their property, and place it in a situation where it would not benefit either themselves or the community, they are just as guilty as the man who steals my property. You may inquire—"What should be done with such a character?" Why, CUT HIM OFF FROM THE CHURCH. *I would disfellowship a man who had received liberally from the Lord, and refused to put it out to usury.* We know this is right.

I recollect well the days brother Grant was telling of, when it was so hard to raise fifty dollars for brother Joseph. I also remember we had a man for trial before the high council, a man who had plenty of money, and refused to loan it, or use it for the advancement of the cause of truth. He would not put his money out to usury. I was going into the council when he was making his plea, and he wept and sobbed. His name was Isaac McWithy, a man about fifty-three years of age. I knew him when he lived on his farm in York State. He told them, in his plea, what he had done for the cause, that he had always been a Christian, and had done so much for the churches, and for the priests, and been so liberal since he had been in this church, which was between three and four years. Some of the brethren said—"Brother McWithy, how much do you suppose you have ever given for the support of the Gospel?" The tears rolled down his cheeks, and he said, "Brethren, I believe I have given away in my lifetime two hundred and fifty dollars." I spake out and said, "If I could not preach as many months each year in this kingdom as you have been years in this Church, and give no more than two hundred and fifty dollars, I should be ashamed of myself."

On one occasion, brother Joseph Young and myself had travelled more than two hours among snow, and in a piercing cold, to preach in his neighborhood one evening. Having had no dinner or supper, we went home with him, and he never asked us to eat a mouthful of supper, though he did muster courage enough to go into the cellar with a little basket, he came up with the tears almost running down his cheeks, and said with some difficulty—"Brethren, have some apples." He held out the basket to us, and when we were about to help ourselves, his niggardly soul made him draw it back again, for fear we should take any. I saw he did not intend us to have any apples, so I put my hand on the basket, and drew it out of his hand, saying—"Come here." I took it on my knees, and invited brother Joseph to eat some apples. He did make out to give us some breakfast in the morning, and even then he got up from the table before we had time to half finish our breakfast, to see if we would not give over eating. Said I—"Never mind, I shall eat what I want before I stop."

I am happy to say, through your trustee in trust, that the latter-day saints, in the capacity of a church and kingdom, do not owe near as much money as they have on hand. A year ago last April conference, we owed over sixty thousand dollars, but we do not now owe a single red cent.

May God bless us, that we may always have enough, and know what to do with what we have, and how to use it for the good of all, for I would not give much for property unless I did know what to do with it.

"MORMONISM."

A DISCOURSE DELIVERED BY PARLEY P. PRATT, IN THE TABERNACLE, GREAT SALT LAKE CITY.

I HOPE the congregation will lend us their undivided attention, and exercise their faith and prayers for those that speak, that the truth may be drawn out to the edification of all.

I always feel diffident to address the assemblies of the people of God, at the seat of the government of the Church, knowing that there are many that can edify and enlighten our minds better than I can. I always feel that I would sooner hear than speak. But nevertheless, I feel it my duty to impart my testimony, and exercise my gift among my brethren, according to my calling; I therefore shall address you for awhile this morning.

There may be many strangers assembled with us at this season of the year; many are passing through this city from different parts of the world. The members of the Church need not complain, if I should address myself to the people as if they were all strangers, on the principles that are sometimes designated " MORMONISM ;" and confine myself to some of the plain, simple, introductory principles of that system. It will refresh the minds of those acquainted with them, and perhaps edify them, and at the same time edify others.

Suppose I were to ask a question this morning, as a stranger, "What is Mormonism?" I suppose it is known to most men at all conversant with principles classed under that name, that it is a nickname, or a name applied to the public, and not used officially by the Church so called. Mormon was a man, a prophet, an author, a compiler, and a writer of a book. Mormon was a teacher of righteousness, holding certain doctrines. The Church of Jesus Christ of latter-day saints are agreed with Mormon, as well as with many other ancient writers, and hold to the same principles; therefore their neighbors have seen fit to call those principles they hold, "MORMONISM." They might as well have called them, *Abrahamism*, *Enochism*, or *Isaiahism;* because the ancient prophets, patriarchs, and apostles, held to the same truths in general terms, only differing in circumstances, in distant countries and ages of the world, and acted upon the same general principles, according to the particular circumstances that surrounded them. But the world, out of all the ancients, have selected one called *Mormon*, and all the principles held by all good inspired men of all ages and countries they have seen fit to sum up, and call "Mormonism." Well, it is as well as anything else, for aught I know; name does not affect the principles.

The word of God, as written in the good old Book, designates the people of God by the name of Saints; which name is almost or quite as ancient, as any writings extant. Saint was spoken of by Enoch long before the flood. The same term was applied to the people of God by the prophets, the psalmist, and by the writers of the New Testament.

Not only was this term applied to saints in ancient

days, but the patriarchs, prophets, and apostles applied it prophetically, speaking of the people of God in the latter days, when the kingdom should be given to the people of God, and the principles of God should bear rule over all the earth. Daniel and the other prophets, in speaking of this subject, always call them the saints of the Most High. They do not call them "Mormonites," Methodists, Presbyterians, Congregationalists, Jews, Pagans, or Mohammedans, nor yet Catholics; but the language of the apostles and prophets is, that the SAINTS of the Most High shall prevail—prevail over the world, establish a true order of government, and, in short, rule the lower world; and that all the nations shall bow to him who is at their head, and to the principles held by them.

Why not this be continued and sustained, Oh ye people of Christendom, and, letting these party names go by the board, and be classed among the things that were in the darker ages, come to the proper and correct Scripture language, and when we speak of the people of God, call them SAINTS OF THE MOST HIGH?

Well, then, such is the name that the Church which I represent, do their business in. As such, they are known on their own records, and on the records of heaven, inasmuch as they are recognized there. But we know what the world mean when they say "Mormonism," and "Mormon." What are the principles called "Mormonism?" You may ask those who profess to be instructors of the people abroad in the States, and elsewhere—and very few of them will give you one correct idea in regard to the doctrines of the latter-day saints. Indeed, they have not informed themselves, but remain in ignorance on the subject; and when they would show others, of course, they cannot inform them correctly on that subject. But you will generally be informed, that "Mormonism" is a new religion, that it is something new under the sun, and of course is an innovation—a kind of trespass on Christianity, on the Bible, or on the good old way. "Oh," say some of the editors that ought to be the most enlightened, and that profess to be, "if Mormonism prevails, Christianity will come down."

Now suppose that we examine, principle by principle, some of the fundamental principles of "Mormonism," and see whether there is one item that is new, or that is in any way an innovation on Christianity.

What is the first start towards an introduction of these principles in this age, and the organization of a people? What is it that first disturbed the world, or any part of it, or called the attention of the people towards it, giving rise to the system now called "Mormonism?" It was the ministration of angels to certain individuals; or in other words, certain individuals in this age enjoyed open visions.

Now we will stop, right at this point; it is called "Mormonism." Let us dwell on it. Is that a new principle? Is it adding something to Christianity, or taking something from it? Do not let our modern notions weigh anything, but come right to the fact of the matter. If Peter the apostle were here to-day, and a person were to relate to him a vision wherein an angel appeared to him and said something to him, would Peter call together the rest of the apostles, and sit in council on that man's head for error? Would they say to that man, "Sir, you have introduced something here in your experience that is derogatory to Christianity, and contrary to the system of religion we have taught, and introduced into the world?" I need not answer this question, neither need I bring Scripture to show what were the teachings and experience of Peter and the rest of the apostles on this subject. The Bible is too common a book, too widely circulated in the world, and the people of the United States, especially, are too well read in its contents to suppose, for a moment, that Peter or the rest of the apostles would condemn a man because he believed in the ministration of angels, because he related an experience wherein he had had a vision of an angel.

Now that was the principle that disturbed this generation, in the commencement of the introduction of that which is now called "Mormonism"—a principle as common in the ancient church as the doctrine of repentance. I will say more—it is a principle that has been common in all dispensations; it is a principle which was had before the flood, and fully enjoyed by the ancient saints, or at least held to by them; a principle that was common among them; not that every man attained to it.

But where can we read, under the government of the patriarchs, before the flood or after it; before Moses or after him; before Christ or after Christ—where can we read in sacred history of a people of God by whom the doctrine of visions and ministering of angels would be discarded, or be considered erroneous? It was common to all dispensations, it was enjoyed by the patriarchs and prophets under the law of Moses, before it and after it, and by the people of God among the ten tribes, and among the Jews. We will carry it still further. It was enjoyed among the Gentiles, before there was a people of God fully organized among them in the days of

Christ. Cornelius had the ministering of angels before he became a member of the Christian Church, or understood there was a crucified and risen Redeemer. He prayed to the living God, and gave alms of such things as he had. He was a good man, and an angel came to him and told him his prayers were heard, and his alms had come up as a memorial before God.

It is astonishing then, to me, that the modern Christian world consider this a new doctrine, an innovation—a trespass on Christianity. No! it is as old as the world, and as common among the true people of God, as His every day dealings with man. We will leave that point, and say, it is the Christian world, and not the latter-day saints, that have a new doctrine, provided they discard that principle.

What next? Why, that man, by vision, the ministering of angels, and by revelation, should be called with a high and holy calling—commissioned with a holy mission to preach, and teach, and warn, and prophesy, and call men to repentance. That was one of the first principles introductory to what is now called "Mormonism" in this age.

Is there anything new about that, anything strange, anything that differs from the patriarchal ages, from the Jewish economy, the Mosaic dispensation, or from the dispensation called Christian? Similar things happened before Moses, in his day, and after his day; and among the prophets, and in different ages. Were not such things common in the days of Jesus Christ, and after that in the days of the apostles? Was not John the Baptist thus commissioned? Was not Jesus thus commissioned. And were not His apostles, elders, and seventies? After his resurrection and ascension into heaven, were not others called, and ordained under the hands of those who were thus commissioned, and called sometimes by visions and revelations directing them to those who were thus commissioned in order to be ordained? That was no new doctrine, no innovation on Christianity, no perversion of the Scriptural system, nor was it anything new, unless you call the old principle new.

Well, then, that the man thus commissioned should call upon others to turn from their sins; and that an individual, a government, a house, a city, a nation, or a world of people should perish unless they did turn from their sins—is that anything new? No. Every one conversant with the Bible will say, that such things took place frequently under all the different dispensations. The heathen were warned in this way. Individuals, households, cities, nations, and the world have to be warned in this way, and especially under the Christian dispensation. So there was a special commission given to the servants of God, to go to all the world, and call upon everybody to repent, or whole nations should become disfranchised, scattered, and millions be destroyed, as for instance the Jews at Jerusalem, because they would not hearken to it. It is nothing new, to cry to all men to repent, and warn different cities and nations of wars coming upon them, or that they will be damned if they do not repent. This is one of the early principles called "Mormonism." Is there anything new in this? Is there anything strange or unscriptural? No; no sensible professing Christian will maintain such a point for a moment.

Suppose that some people should hearken, when the ministering of angels takes place. Among many men one certain man is commissioned by revelation to preach the Gospel, and cry repentance. Suppose that some persons should hearken and repent, and he should take them and walk down to the water, and bury them in the water in the name of the Father, and of the Son, and of the Holy Ghost, and raise them again out of the water, to represent the death and burial of Jesus Christ, and his resurrection from the dead; and to represent the faith of the individual thus ministered to, that he does believe in Jesus Christ, that he died, and that he did rise from the dead, and that he, the individual, does put his trust and confidence in him for the remission of sins and eternal life—is that anything new? Would that be new to Peter? Suppose some person was to relate before Peter and Paul to-day, and the Christians with them, that lived when they lived—suppose they were all present, and this person told them that a man came along preaching repentance, and he called upon us to believe in Jesus Christ, and we did so, believing their testimony, and they took us and buried us in water, and raised us again out the water unto newness of life—would Peter or John blame him? Would Paul say, "It is something new?" Or would he say, "Brother, thousands of us received the very same thing in ancient days?"

The Catholic Church profess to be the true church— the ground and pillar of the truth, handed down by regular succession from the ancient church, of which they are still members; and their priesthood and apostles are now of the very same church which the New Testament calls the true Church at Rome. These Roman Catholics of modern times profess to be members of that very same church that Paul wrote that epistle to. If they are, I will show you to demonstration, if the

Scriptures be true, that this doctrine called "Mormonism" is not a new doctrine, Paul, writing to that church, of which they profess to be members, says, Know ye not, brethren, ye Romans, that as many of you as have been baptized into Christ have been baptized into his death, being buried with him by baptism into death, that like as Christ rose from the dead, even so ye may walk in newness of life? Now this epistle containing this doctrine was written by Paul to the Church at Rome, and which these modern people called Roman Catholics profess to be members of. If they are what they profess to be, every one of them have been buried with Christ in baptism, and have risen again to newness of life. We will, however, leave them to describe whether that is really the case, or whether they are contented to sprinkle a few drops of water on an infant's face and call that a burial! Paul said that was a principle of the true Church of Rome that had been buried with Christ by baptism into death, and had risen to newness of life. Have these modern Roman Catholics gone forward repenting of their sins, and been buried in water, in the likeness of Jesus Christ according to this pattern? If they have not, they are a spurious Church of Rome, and not real. Therefore, if they be the real Church of Rome, it will be no new thing to them when the latter-day saints inform them upon being buried with Christ in the likeness of his death, etc. If this is a new doctrine to them, they had better be looking about them to see if they have not got up a counterfeit Church of Rome, for Paul knew of only one, and the members of it were all buried with Christ in baptism.

If 500 persons here were to say they came repenting of their sins, and went down and were buried in the waters of baptism, and had risen again to walk in newness of life, Paul would say, if he were here, "It is just what we used to do in ancient times; and I wrote to the Church of Rome, telling them that as many as were baptized into Christ were baptized into his death, buried with him by baptism into death," etc.

Now if this doctrine is new to the Church of Rome, then that is that church, that priesthood, and those members that have introduced something new, who are departing from the old Christian religion, and not the "Mormons."

This reasoning applies just the same to the Church of England. They have just as good a right to have a Church in England as anywhere else—to have a national Church of England by law established, but if they are a true Church of God, all of them have been buried with Christ in baptism, etc., or the apostle must have been mistaken, or there are two different kinds of Gospel.

Now if I were speaking to the state Church of England, or the state churches of the Catholic world, I would tell them in the name of the Lord Jesus Christ to repent of their new doctrine, and come back to the old standard spoken of by the Apostle, when he says, "though we or an angel from heaven preach any other Gospel unto you than that which we have preached unto you, let him be accursed," etc.

I need not go through with this same application upon the Lutherans, upon the Presbyterians, upon the Methodists, and others, for all these people sprinkle infants; for the principle once carried out will apply to the whole. If they are Christians according to the doctrine of the ancient church, they hold the doctrine of the apostles, they have repented of their sins, after believing on the Lord Jesus Christ, and have been BURIED with Christ by baptism into death, etc. If not, they may judge themselves, for I will not judge them. If they have got a new doctrine, different from that believed by the apostles, and the latter-day saints have got the old one, why not say, then, "If sectarianism prevails, Christianity, as held by the Mormons will be in danger," instead of saying the opposite? Why not turn the thing right about? If we have no one new principle in our religion, why are we considered innovators, and opposed to Christianity? And why is Christianity in the world in danger if "Mormonism" prevails? It is because that floating Christianity, called so by the world, is a spurious one; they have departed from the doctrine of the apostles. Then, I ask again, why say, "If Mormonism prevails, Christianity is in danger?" for if it is a false Christianity, the quicker it falls the better.

We have examined three general principles, to see if there is anything new in "Mormonism." First, the ministering of angels. Second, the commission of ministers, apostles, prophets, and elders to administer in holy things, by revelation and the authority of heaven. Third, that all those that hear them, believe their words, and repent of their sins, shall go down into the waters of baptism, and be immersed or buried in the name of the Father, and of the Son, and of the Holy Ghost, and thus show that they do believe in a crucified and risen Redeemer, and in the remission of sins through His name. So far, I think, we have fairly stated some of the first principles of what the world calls "Mormonism;" and every one who has heard us, must decide that there is nothing new in these principles, but rather, that those who have departed from them, are justly chargeable with introducing new things, and innovations on Christianity.

Now suppose that one, two, or a dozen, or a hundred thousand, or even millions of individuals thus baptized, should all come together, in their several congregations, and should unite in earnest prayer, and a man commissioned in the ministry of Jesus Christ should rise and lay his hands on them, praying the Almighty God to give the Holy Spirit, and it be given as in days of old, and he confirms that promise upon them according to the pattern in the New Testament—would that be something new? Would it be an innovation upon Christianity? Would it be right to say "this is Mormonism, come to do away with Christianity?" Why, no! Every sensible man at all acquainted with the Holy Scriptures, would laugh at the idea. If the ancient saints were here, they would tell you it was their ancient manner; they would ask you if you had not read over their history, which describes how the Holy Spirit was administered in days of old. Every man who has read the Bible, knows it.

Well, then, the different sections of what is called Christianity, never do this, and call it something new. When the "Mormons" do it, they are at once charged with innovation; and yet we have not got anything new in that respect, but simply a restoration of that which was. They are the persons chargeable with new doctrine, and not the latter-day saints.

Well, then, suppose that after this ordinance, the Holy Spirit falls upon these congregations, or upon these individuals thus baptized and confirmed, and fills them, and enlightens their minds, and bears testimony to them of the truth which they have received, and confirms them in the faith of it, and fills them with the spirit of utterance and prayer, and with gifts whereby they prophesy, or speak in tongues, lay hands on the sick and they recover, in the name of Jesus, or whereby they are filled with the spirit of any gift, renewed in their utterance, strengthened in their powers of intellect, so as to be able to speak with eloquence to the edification of others by the word of wisdom, knowledge and prophecy; or peradventure some one, two, or three of them have a heavenly vision, and happen to relate it—is this something new? Are these things an innovation on Christianity?

Let the apostles of the ancient church come up now, and be judges, not these innovators. Oh yes, saints of ancient days, are these things new to you? "No," they reply, " but just exactly what we used to have among us; and you who have read the New Testament know it is so." If this, then, is Mormonism, it is nothing new, but simply that which should have been in the world in order to constitute true Christianity.

Now suppose, after all these have been established, the people organize on them; and that in the enjoyment and cultivation of them, this people unite their efforts, both temporally and spiritually, to build up themselves as a people, and each other as individuals, in righteousness upon the earth; and the Spirit of the Lord God into which they were all baptized, should make them very great in union—in union of effect, in counsel, in operation, in fellowship, in temporal things in a great measure, and in spiritual things, by which they are all of one heart and mind to a great degree, and growing in it every day—is this something new, because it is "Mormonism?" Or is this the very doctrine which was inculcated in days of old by the apostles of Jesus Christ?

It was the main object for which the Holy Spirit was given, that they might all grow up in union, in fellowship, in co-operation, in holiness in the Lord. No man who has read the New Testament, will say this is NEW, when we say that the great object of the Gospel is, that we may all become one in Christ Jesus—one in knowledge, and in the love and practice of the peaceable things of God. Is it anything new? No. Well, it is a part of what the world calls "MORMONISM;" and I would to God it was more perfected among this people than it is.

If any one of these principles in practice, should prevail over the whole world, it would be nothing new; but the world only hold this last as a theory; as to the practice of it, they are strangers.

We have examined five or six general principles, called "Mormonism," and found nothing new in them. "But," says one, "I heard you had got a new Bible; that is certainly an innovation." But stop; suppose, on inquiry, you become as much surprised and disappointed as many have who have asked for a "Mormon Bible," and when we have presented them with one, behold it is King James' translation of the Scriptures, the standard we read, containing the covenants, predictions, and hopes of the ancients, and the doctrines of Jesus Christ, just as we believe them, and hope for their fulfillment. Is that anything new?

"Well, if you have not a new Bible, you have certainly got a new book." Is that any thing strange? Have not other societies got new books? The Church of England have not only the Scriptures, but the book of Common Prayer, and the time was when they did not have such a book, therefore when they made that, it was something new. They are not alone in that, however, for the Methodists have a new book called the "Methodist's Discipline." One hundred and twenty years ago there was no such thing in existence. If having a new book be an innovation, then all are guilty of it as well as the "Mormons."

"But those other people do not profess that their books are inspired, and we have learned that you have a book that you believe is inspired. What is it, any how?" This is all a fact, and if it is wrong we will cheerfully plead guilty. We have got another book besides the Bible, that was an ancient book, and profess that it was inspired, and was written by prophets, and men that enjoyed the ministering of angels, more or less of them, and had communion with the heavens, and the spirit of prophecy. And moreover, we profess that this ancient book was restored to the knowledge of the modern world by inspiration, and the ministering of angels. Is that something new? It may be new to the world in its history, and in its bearings; in that respect it may be new to them; but suppose, after all, it should contain no new doctrine, no new principle, no new prophecy, that is differing from or doing away that which is already extant in the Bible? Well, then, I do not say that it would be a new doctrine. Men had books revealed in the days of old.

"If it is no new doctrine, and if its predictions do not differ from those contained in the old and new Testaments, what is the use of it?" The same question was investigated in ancient times. A great conqueror had taken possession of an ancient library, when there were no printing presses, containing one hundred thousand volumes, all in manuscript, comprising more history than was in any library extant in the ancient world. The conqueror was a Mahommedan. He wrote to the head of the department to know what he would do with this library. It was invaluable in its cost and intrinsic worth. "What shall I do with it?" The reply was, "If it agrees with the Koran, we have no use for it; and if it does not agree with the Koran, it is false anyhow; so in either case burn it."

"Now if these latter-day saints have a book extant among them, and it agrees with the Bible, there is no kind of use for it," says the opposer, "for the Bible contains all that is necessary; if it does not agree with the Bible, it is false anyhow; so in either case burn it." This was a principle of Mahommedanism, and may be a principle of what is called modern Christianity. I hope not, however.

"What is the use of the book in question, anyhow?" Why, in the first place, it differs in its history from the Bible. The Bible is a history of things that took place in Asia principally, and a little of what took place in Europe and Africa. The book of Mormon is a history of things in another hemisphere. The one book is the ancient history of the eastern hemisphere, in part; and the other is a history of the western hemisphere, in part. Shall we say, because we have the history of one part of the world, that the history of the other part of the world is good for nothing? Could the rulers of nations realize that fact, and could they only have a copy in their libraries at the cost of $100,000, they would appropriate it for this history of the western hemisphere.

Discredit it as you will, we have it in genuineness

and in truth, written by the ancient prophets that lived upon this land, and revealed in modern times by the ministering of angels, and inspiration from the Almighty. It is in the world, and the world cannot get it out of the world. It is in the world in six or seven languages of Europe. It is as important in its history as the Bible, and it is just as interesting and as necessary for men to get an understanding of the ancient history of America, as it is for them to get an understanding of the history of Asia.

"But are the merits of history all that it is good for?" It is good in doctrine also. If two or more writers, one living in Asia, and the other in America, and cotemporary, have the same doctrine revealed to them, and both bear record of the same plan of salvation, who is he that shall say that the record of one is of no worth?

Is it not a satisfaction to sit down and read that a country far removed from Bible scenes, from that part of the stage on which figured the patriarchs of old, with Moses and the Jewish prophets, John the Baptist, Jesus Christ, and the apostles, was also the theatre of revelation, prophecy, visions, angels, of the ministration of the doctrine of Christ, of the organization and government of his true church? that there too were angels, that there too were apostles, that there too was the word of God, that there too faith came by hearing, and salvation by faith? Shall we say that such things and such good news are worth nothing, when that very news corroborates the song of the heavenly hosts, when they declared to the shepherds of Judea, in joyful songs, that they brought glad tidings of great joy, that should be to all people! And here comes a book informing us that these glad tidings were also to another hemisphere at the same time.

Now, stop a moment, and let us reason. Suppose yourself an angel of God at that time, full of benevolence, full of joy, full of a soul-inspiring hope, full of charity for poor, ignorant, perishing mortals, and you felt so full of poetry, and song, and gladness, that you could scarcely hold your peace. Suppose you had a bird's eye view of our little, dark, benighted world, by soaring above it, and in a moment you could light down upon any part of it. You come to Palestine, in Asia; that part of the globe is rolling under your feet; you visit it, and sing to the shepherds the glorious tidings of great joy, which shall be to all people: "for unto you is born this day in the city of David, a Saviour which is Christ the Lord." The earth rolls on about half way round, you look down again with a bird's eye view, and you discover the western hemisphere, and it is full of people; I wonder whether your soul would still swell with the same glad tidings—or would your charity have become exhausted? Would you not fly and declare these glad tidings to them also, and sing them a song of joy, and tell them what day the Saviour was born, that would reach their case as well as the case of those who dwelt upon the continent of Asia? "Yes," you reply, "if I were an angel, and had liberty to tell these glad tidings, I would never tell them to one part of the earth and go to sleep there, while the other part rolled under my feet unnoticed."

Were those angels commissioned and endowed to bear glad tidings to ALL PEOPLE, that the Saviour was born? I say that the choir of angels which sang that song, had full liberty, not only to tell the plan of salvation to chosen vessels of the Lord in one country, but also to another country—not only that the Saviour was born, in general terms, but the place where, and the time when, he was born. These were the tidings, "Go to all people." An angel must be a limited being, or be very ignorant in geographical knowledge, or partake largely of sectarian feelings of heart, to bear such tidings to one half of the globe, and not to the other.

I knew an infidel once, that did not believe in the Christian religion, nor in the New Testament, nor in the Saviour of the world. I asked him why he did not believe this. "Because," says he, "according to the New Testament the manifestation of such an important affair was so limited. Here was half of the world, according to the New Testament, that never heard of it. A message so important should have been made more public.'

"Well," said I, "if I will produce you a record, and a history, as well authenticated as the New Testament, showing that angels, the risen Saviour, holy inspired prophets and apostles, ministered in the western hemisphere, and preached the gospel to every creature, and handed it down, to ages, will you then believe?" "Yes," he answered, "I will." I presented him the Book of Mormon, which he perused. I inquired if he now believed. "Yes." he said, "I do;" and he has lived a Christian until now, for ought I know. I have seen him in this congregation, and he may be here today. His name is Alger.

What objection have you to the hope of eternal life being as widely developed as the ravages of death, sorrow, and mourning? What objection have you to the angels of God, apostles of God, the Son of God, or to the Holy Spirit of prophecy being poured out in more countries than one? You may say the keys of the Gospel were given to the Jewish apostles, but they were so far off as not to be able to reach the western hemisphere, even if they had had a knowledge of it. Were there ships and steam vessels to bear them to this country? No. Was there any communication kept up, or was this country known to them? No. But the waves, and winds, and elements, and the great depths that intervened, even the unexplored ocean, said to the ancient apostles, "Thus far shall ye go, and no further." This ocean, however, was no barrier to the fleet-footed angel of God, to the risen Jesus, and to immortal man. They could come to this hemisphere, and reveal the things of heaven to the people, and could rejoice in the same glad tidings, whether it was here or in Jerusalem, or if it were in the uttermost parts of the earth.

Though Peter was crucified at Rome, and Paul suffered in the same manner; though saints of the Most High were slaughtered by thousands and tens of thousands, and bled at the feet of Roman altars; yet a crucified and risen Redeemer, angels of God, and the Holy Spirit of truth that fills all things, were not thus curtailed and limited, but could minister truth to the uttermost bounds of the universe of God, where intelligences were mourning in darkness; wherever the ravages of death had spread sorrow, wherever there was a broken heart to be bound up, or wherever there was a despairing mortal to be inspired with hope, they could go and tell the glad tidings of life and salvation. The Book of Mormon says they did come to this continent. It is a history of their coming, and contains the doctrine taught to the people here by the risen Jesus, and by his predecessors. In short, the doctrine taught and practised in ancient America is there portrayed, together with the history of the people.

Again, is this book of no interest with regard to the prophetic value? It reveals many things not noticed by the Jewish prophets. Did the old prophets touch upon every item that pertains to man in other countries? No, they did not, only in general terms together with the rest of the world. These other prophets portrayed many things not in their book, though agreeing with it as far as it goes, but touching events on which their book is silent.

Has any person any cause to say that there has not been a multiplicity of revelations, testimony, prophecy, history and doctrine, developed in various countries by the same Spirit of God, and by angels? And is not all this of great worth, to compare, in order to blend it together, that we may see more clearly the principles of the doctrines of salvation, and understand prophecy more extensively, especially in an age when the mind has been obscured by priestcraft.

If these are the principles of "Mormonism," where can you point out an innovation on Christianity? But is this all? No, this is not all, and I shall not tell it all today. I do not know it all yet. I have been twenty-three years learning "Mormonism," and I know but little of it. If any one expects to learn all the doctrines of "Mormonism," he must learn more than twenty-three years. For be it known unto you all, that "Mormonism," instead of being confined to a few dogmas or general truths, opens the flood-gates of all truth and knowledge, and teaches mankind to retain all the truth they can already comprehend, and comprehend as much more as they can all the time.

"Have you not other books?" Yes, we have histories and compilations of the dealings of God with us as a people. We keep a record, if you must know, not only individually some of us, but as a church, as a body, or community. We have revelation penned, revelations and visions penned, we have revelation and prophecy penned, we have knowledge penned, we have knowledge and principle penned, we have principle and history penned; the history comprising but a small portion, such as can be written, revealed to us latter-day saints, and practised upon; so that our modern books are like the ancient books—a mixture of revelation, prophecy, history, and doctrine. Has any person any objections to this? I ask, should an angel administer to this or that man, or suppose an open vision was manifested to him, revealing many precious truths, would he not be a simpleton not to write it? If the power of God, and the ministering of God, and the visions of the Almighty are extant in the world, these will be written. The practical part of history will be written, for if all were written, the world would not contain the books. The ancient apostles and prophets wrote a few of the items revealed to them, and a history of the practical workings of the system over which they presided. Do we differ from them? No.

"Well," says one, "to be plain with you, Mr. Speaker, we have been taught to believe that the one book, called the Bible, contains all the revelations that God ever revealed to man, therefore it is an innovation to offer anything else to the world as a revelation." This is a tradition of your own, so I have nothing to do with it. The Bible never taught that to you, nor angels, neither did any minister of God ever teach it to you; and if it is a modern sectarian tradition, it is calculated to bind men into a cast iron creed, and the sooner you break the fetters the better; burst them asunder, and come out into liberty and freedom, and know and understand that there is no such doctrine in the broad principles of eternal truth, that heaven is full of knowledge, and the earth ought to be full of prophets, heaven and earth full of angels, and both full of inspiration; and if the inhabitants of all the worlds of the universe were scribes, every blade of grass a pen, and every ocean ink, they could not write all the doings of the Almighty, of His servants, and of His angels. If I were to live for millions of years to come, and then millions of millions more, I expect there would always be some being ready to reveal something new, and somebody would write it. The art of writing will never cease. We may not have pens and ink, but we may have something better. Suffice it to say, that the arts and sciences will not come to an end, yet man may have been traditionated to believe that one small book contains all that God ever said or did. Such persons are to be pitied, and not to be reasoned with.

What is "Mormonism?" It is a restoration by new revelation, by the authorities of heaven, by the ministration of angels, by the ordination of prophets and apostles, and ministers or elders, by their testimony and ministry on the earth, by the organization of saints, by the administration of ordinances, by the operations of the Holy Spirit; it is a restoration of these ancient principles revealed from heaven, for the government of man.

Says one, "You have said you are not going to tell the whole system to-day." I do not know it all, and I shall not state the half I do know. What I have said are a few every day items, a few of the first principles of the Gospel of Christ, as believed and practised by the "Mormons."

I will tell one more before I close. "Your marriages," says the objector, "are founded upon principles entirely new, and different from the Christian world." I say, without any hesitancy, I defy the world to establish that assertion. I say our marriage relations are nothing new at all. There is no man, or set of men, or nation of men, where the Bible is extant, and they are readers, but what know that the institutions of marriage contained in the Bible, and the organization of families, differ widely from modern Christianity. We differ from modern Christianity, but not from the Bible. Patriarchs of the remotest ages, that obeyed the Lord God in regard to their marriages and family organizations, have not disagreed with us, nor we with them, so far as we and they have obeyed the law of God. If there is any difference at all, it was more developed among them than it is among us, we being in our infancy. If it should happen to be, that the whole modern world differ from the Bible—have done away with the law of God, and we have come in contact with them, instead of with the word of God, then the boot is on the other foot, and in reality what is said to us applies to them. It is like the farmer and the lawyer. A certain farmer came to a neighboring lawyer, and frankly confessed that his bull had had the misfortune to kill one of his (the lawyer's) oxen. The lawyer replied, "Thou art a very honest fellow, and will not think it wrong that I have one of thy oxen in return." "But," said the farmer, "I am mistaken, it was thy bull that killed my ox." "O," replied the lawyer, "that alters the case, and if—if—i-f—."

Now, then, if it is the whole Christian world, from Catholicism down to the latest of her daughters, that have made void the law of God, and trampled under foot the institutions of heaven, the holy principles of matrimony and family government, and have made them void also, by their traditions, and introduced that which God never did; and "Mormonism" has restored the law of God, in theory and practice, then it is the so-called Christian world, and not us, that are wrong. Whether it regards family organization, the law of God, patriarchal government, ordinances, principles and prophecy, I know of nothing new, or of nothing wherein we are innovators.

As I said before, and I am able to maintain it when called upon, "Mormonism" is a system which was understood and enjoyed by the ancients, and restored unto us by revelation. And if carried out, what will it do? It will simply fulfill the sayings of the prophets, both ancient and modern, put down all wickedness, abuse, proscription, misrule, oppression, ignorance, darkness, and tyranny, and restore mankind to righteousness, truth, liberty, law and government, in which the Lord's will will be done on the earth as it is in heaven. That is what "Mormonism" will do, when carried out.

May God bless you all. Amen.

LEGITIMACY AND ILLEGITIMACY.

A SERMON DELIVERED BY ELDER JOHN TAYLOR, AT THE GENERAL CONFERENCE, IN THE TABERNACLE, GREAT SALT LAKE CITY.

It rejoices my heart to hear the principles that have been advanced this day by our President, because they have their foundation in truth, and based upon the principles of equity, and are calculated to promote the happiness, well-being, exaltation and glory of man, in time, and throughout eternity. They lead us back into eternity: they existed with us there, and in all the various stages of man's existence they are calculated to elevate and ennoble him, and place him in a proper position before God, angels and men. They will put him in possession of his legitimate right, save him from the grasp of the adversary, from every subtle strategem of the powers of darkness, and place him in his proper station in time and in eternity.

I have been much pleased with and edified by the remarks that have been made upon this stand during the conference. Wisdom has been displayed in them; from them the intelligence of heaven have beamed forth, the mysteries of eternity have been spread before our minds, and we have had a view of heavenly things, that has filled our hearts with joy and our mouths with praise.

It has made us feel as though we were upon the threshold of eternity; as though we were eternal beings, and had to do with eternal things; as though the things of this world were short, fleeting and evanescent, not worthy of a thought when compared with those things that are calculated to exalt and ennoble us in time and in eternity.

The principles of justice, righteousness, and truth, which have an endless duration, can alone satisfy the capacious desires of the immortal soul. We may amuse ourselves like children do at play, or engage in the frivolities of the dance. We may take our little enjoyments in our social assemblies, but when the *man* comes to reflect, when the *Saint of God* considers, and the visions of eternity are open to his view, and the unalterable purposes of God are developed to his mind—when he contemplates his true position before God, angels and men, then he soars above the things of time and sense, and bursts the cords that bind him to earthly objects; he contemplates God and his own destiny in the economy of heaven, and rejoices in a blooming hope of an immortal glory.

Such have been some of our feelings, while our minds have been carried away from the things of earth to contemplate the things with which eternal beings are associated, and the glories that await us in the everlasting mansions of the Gods.

The principles that we have to do with, then, are eternal, and not simply to play a game upon the checker of mortality, on which people can win and lose for the time being. We have to do with that which shall continue

> "While life, and thought and being last,
> Or immortality endures."

We seek not to build our hopes upon things that are evanescent, fleeting and transitory.

It is not he that can play the best game at checkers, that can take the most advantage of his neighbor, that can grasp the most earthly good, or that can put himself in possession of anything his heart desires pertaining to time, that is the most happy; but it is *he* who does to that which will *last, live* and *continue* to abide with him while "*immortality endures*," and still be on the increase, worlds without end.

If we can possess principles of this kind, then we are safe, everything else amounts to an illusion or a delusion, which cannot satisfy the desires of the mind, but, as the prophet says, it is like a thirsty man who dreams he is drinking, but when he awakes, he is faint, and his soul is thirsty; he dreams that he is eating, and when he awakes his soul is empty. This is the true situation of all men who are without God in the world; and nothing but a knowledge of eternal principles, of eternal laws, of eternal governments, of eternal justice and equity, and of eternal truth, can put us right, and satiate the appetite of the immortal soul.

If we make not a just estimate of these things, it is in vain that we attempt to say, "Lord, Lord," because we do not the things which he says. Every thing associated with the Gospel of salvation is eternal, for it existed before the "morning stars sang together for joy," or this world rolled into existence. It existed then, just as it exists now with us, and it will exist the same when time with us is no more. It is an eternal principle, and every thing associated with it is everlasting. It is like the priesthood of the Son of God, "without beginning of days or end of years." It lives and abides for ever.

If there is any principle that is not eternal, it is not a principle of the gospel of life and salvation.

There are many changes and shifting scenes that may influence the position of mankind, under different circumstances, in this state of mortality; but they cannot influence or change the gospel of the Son of God, or the eternal truths of heaven; they remain unchangeable; as it is said very properly by the Church of England, in one of their homilies, "as it was in the beginning, is now, and ever shall be, worlds without end." If nothing else they say is true, that is, and I can say amen to it with all my heart. All true principles are right, and if properly understood and appreciated by the human family, to them they are a fountain of eternal good.

The principle of "heirship," which President Young preached about to-day, is a principle that is founded on eternal justice, equity and truth. It is a principle that emanated from God. As was said by some of our brethen this morning, there may be circumstances arise in this world to pervert for a season the order of God, to change the designs of the Most High, apparently, for the time being, yet they will ultimately roll back into their proper place—justice will have its place, and so will mercy, and every man and woman will yet stand in their true position before God. If we understand ourselves correctly, we must look upon ourselves as eternal beings, and upon God as our Father, for we have been taught when we prayed to say, "Our Father, which art in heaven, hallowed be thy name." We have fathers in the flesh, and we do them reverence, how much more shall we be in subjection to the Father of Spirits and life. I need not enter into any proof in relation to this, for it is well understood by the Saints that God is the

Father of our spirits, and that when we go back into His presence, we shall know Him, as we have known our earthly parents. We are taught to approach Him as we would an earthly parent, to ask of Him such blessings as we need; and He has said, "If a son ask bread of his father shall he give him a stone, or if he ask for fish, a scorpion? If ye then, being evil, know how to give good gifts unto your children, how much more will your Heavenly Father give His holy Spirit to them that ask Him."

We have a Father, then, who is in heaven. He has placed us on this earth for some purpose. We found ourselves in posession of bodies, mental faculties, and reasoning powers. In a word, we found ourselves intelligent beings, with minds capable of recalling the past and launching into the unborn future with lightning speed; and were it not for this earthly tabernacle, this tenement of clay, they would soar aloft and contemplate the unveiled purposes of Jehovah in the mansions of the redeemed. We found ourselves here with minds capable of all this and more. God, who has ordained all things from before the foundation of the world, is our Father. He placed us here to fulfill His wise and unerring counsels, that we might magnify our calling, honor our God, obtain an exaltation, and be placed in a more glorious, exalted and dignified position than it would have been possible for us to enjoy if we had never taken upon us these bodies. This is my faith; it is the faith of this people.

I have no complaints to make about our father Adam eating the forbidden fruit, as some have, for I do not know but any of us would have done the same. I find myself here in the midst of the creatures of God, and it is for me to make use of the intelligence God has given me, and not condescend to anything that is low, mean, grovelling and degrading—to anything that is calculated to debase the immortal mind of man, but to follow after things that are in their nature calculated to exalt, ennoble and dignify, that I may stand in my true position before God, angels and men, and arise to take my seat among the Gods of eternity.

We will now come to the principles of *legitimacy*, which was the text given out this morning—to our rights, privileges, priesthoods, authorities, powers, dominions, etc., etc. And as some of us are Scriptorians, and all profess to believe the Bible, I feel inclined to quote a text from it. Paul, when speaking of Jesus Christ, gives us to understand that he is the first-born of every creature, for by him were all things made that were made, and to him pertains all things; he is the head of all things, he created all things, whether visible or invisible, whether they be principalities, powers, thrones, dominions; all things were created by him and for him, and without him was not anything made that was made. If all things were created by him and for him, this world on which we stand must have been created by him and for him; if so, he is its legitimate, its rightful owner and proprietor; its lawful sovereign and ruler. We will begin with him, then, in the first place, in treating on the subject of legitimacy.

But has he had the dominion over all nations, kindreds, peoples and tongues? Have they bowed to his sceptre, and acknowledged his sway? Have all people rendered obedience to his laws, and submitted to his guidance? *Echo* answers "No!" Has there ever been a kingdom, a government, a nation, a power, or a dominion in this world that has yielded obedience to him in all things? Can you point out one?

We read of the Jews who where a nation that submitted only in part to his authority, for they rebelled against his laws, and were placed under a schoolmaster until the Messiah should come. We read also, in the Book of Mormon, of some Nephites that dwelt upon this land, who kept the commandments of God, and perhaps were more pure than any other nation that history gives an account of. But, with these exceptions, the nations, kingdoms, powers, and dominions of the world have not been subject to the law, dominion, rule or authority of God; but, as it is expressed by one of the ancients, the prince and power of the air, the god of this world has ruled in the hearts of the children of disobedience, and led them captive at his own will. Where is the historian, the man acquainted with ancient lore, who can point me out one government, nation, power or dominion, that has been subject to the rule of God, to the dominion of Jesus Christ, with the exception of those Jews and Nephites which I have referred to? If there has been any such nation, the history of it has escaped my notice. I have never been able to obtain such information.

What has been the position of the world for generations past? They have been governed by rulers not appointed by God; if they were appointed by Him, it was merely as a scourge to the people for their wickedness, or for temporary rulers in the absence of those whose right it was to govern. They had not the legitimate rule, priesthood, and authority of God on the earth, to act as his representatives in regulating and presiding over the affairs of His kingdom.

Perhaps it may be well, at this stage of my remarks, to give you a short explanation of my ideas of government, legitimacy, or priesthood, if you please. The question, "What is priesthood?" has often been asked of me. I answer, it is the rule and government of God, whether on earth, or in the heavens; and it is the only legitimate power, the only authority that is acknowledged by Him to rule and regulate the affairs of His kingdom. When every wrong thing shall be put right, and all usurpers shall be put down, when he whose right it is to reign shall take the dominion, then nothing but the priesthood will bear rule; it alone will sway the sceptre of authority in heaven and on earth, for this is the legitimacy of God.

In the absence of this, what has been the position of nations? You have made yourselves acquainted with the political structure and the political intrigues of earthly kingdoms; I ask, from whence did they obtain their power? Did they get it from God? Go to the history of Europe, if you please, and examine how the rulers of those nations obtained their authority. Depending upon history for our information, we say those nations have been founded by the sword. If we trace the pages of history still further back to the first nation that existed, still we find that it was founded upon the same principle. Then follow the various revolutions and changes that took place among the subsequent nations and powers, from the Babylonians through the Medo-Persians, Grecians, Romans, and from that power to all the other powers of Europe, Asia and Africa, of which we have any knowledge: and if we look to America from the first discoveries by Columbus to the present time, where are now the original proprietors of the soil? Go to any power that has existed upon this earth, and you will find that earthly government, earthly rule and dominion, have been obtained by the sword. It was the sword of men that first put them in possession of this power. They have walked up to their thrones through rivers of blood, through the clotted gore and the groans of the dying, and through the tears and lamentations of bereaved widows and helpless orphans; and hence the common saying is "thrones won by blood, by blood must be maintained." By the same principle that they have been put in possession of territory, have they thought to sustain themselves—the same violence, the same fraud and the same oppression have been made use of to sustain their illegitimacy.

Some of these powers, dominions, governments and rulers, have had in their possession the laws of God, and the admonitions of Jesus Christ; and what have they done to his servants in different ages of the world, when he has sent them unto them? This question I need not stop to answer, for you are already made too familiar with it. This, then, is the position of the world. Authority, dominion, rule, government has been obtained by fraud, and consequently is not legitimate. They say much about the ordination of kings, and their being anointed by the grace of God, etc. What think you of a murderer slaying hundreds and thousands of his fellow-creatures because he has the power, and while his sword is yet reeking with human blood, having a priest in sacerdotal robes to anoint him to the kingship? They have done it. What think you of the cries of the widows, the tears of the orphans, and the groans of the dying, mingling with the prayers and blessings of the priest upon the head of the murderer of their husbands and their fathers?

It is impossible that there can be any legitimate rule, government, power or authority, under the face of the heavens, except that which is connected with the kingdom of God, which is established by new revelation from heaven.

In a conversation with some of our modern reformers in France, one of their leaders said, "I think you will not succeed very well in disseminating the principles of your religion in France." I replied, "you have been seeking to accomplish something, for generations, with your philosophy, your philanthropic societies, and your ideas of moral reform, but have failed; while we have not been seeking to accomplish the thing that you have, particularly, and yet have accomplished it." We began with the power of God, with the government of heaven, and with acknowledging His hand in all things; and God has sustained us, blessed and upheld us to the present time; and it is the only government, rule and dominion under the heavens that will acknowledge His authority.

Brethren, if any of you doubt it, go into some of those nations, and get yourself introduced into the presence of their kings and rulers, and say, "Thus saith the Lord God." They would at once denounce you as a madman, and straightway order you into prison. What is the matter? They do not acknowledge the legitimacy, the rule and government of God, nor will they inquire into them. They receive not their authority from Him. Nations honor their kings, but they do not honor the authority of their God in any instance, neither have they from the first man-made government to the present time. If there has been such a nation, or if there is at this time such a government, it is a thing of which I am ignorant.

The kings and potentates of the world profess to be anointed by the grace of God. But the priests who anoint them have *no authority* to do it. No person has authority to anoint a king or administer in one of the least of God's ordinances, except he is legally called and ordained of God to that power; and how can a man be called of God to administer in His name, that does not acknowledge the gift of prophecy to be the right of the children of God in all ages? It is impossible. These men have been grasping after power, and for this they have laid waste nations and destroyed countries. Some of them possessed it for awhile, and others were on the eve of getting it when they were cut off, and down they went. What became of them afterwards? Isaiah in vision saw the kings of the earth gathered together as prisoners in a pit, and after many days they were to be visited.

Having said so much in relation to other governors and governments, we will now notice the difference between them and Abraham of old. Abraham was a man who contended for the true and legitimate authority. God promised to him, and to his seed after him, the land of Canaan for their possession. "The Lord said unto Abraham, after that Lot was separated from him, lift up now thine eyes, and look from the place where thou art northward, and southward, and eastward, and westward; for all the land which thou seest, to thee will I give it, and to thy seed for ever." What did Stephen say generations afterwards? That God "gave him none inheritance in it, no, not so much as to set his foot on;

yet he promised that he would give it to him for a possession, and to his seed after him, when as yet he had no child." Ezekiel's vision of the dry bones explains this seeming contradiction. The Lord said to him, "Son of man, can these bones live?" etc. Who are they? We are told, in the same chapter, that they are the whole house of Israel, and that they shall come out of their graves, bone come to its bone, and sinew to sinew, and flesh come upon them, and they shall become a living army before God, and they shall inherit the land which was given to them and their fathers before them. The measuring line shall again go forth upon those lands, and mark out the possessions belonging to the tribes of Israel.

Abraham was a man who dared fear God, and do honor to His authority, which was legitimate. God tried and proved him, the same as He has tried many of us, and felt after his heart-strings, and twisted them round. When He had tried him to the utmost, He swore by Himself, because He could swear by no greater, saying, "That in blessing I will bless thee, and in multiplying I will multiply thy seed." "And in thy seed shall all the nations of the earth be blessed." Abraham obtained his dominion by legitimate authority; his priesthood was obtained from God; his authority was that which is associated with the everlasting Gospel, which was, and is, and is to come, that liveth and abideth for ever. And the promises made to him will rest upon him and his posterity, through every subsequent period of time, until the final winding up scene of all things. Will he ever obtain them? Yes. For we are eternal beings, and I am now talking as though we were in eternity. We shall wake up in the morning of the resurrection, attain to all the blessings which have been promised to us, and strike hands with Abraham, and see him inherit the promises. Abraham and all his children will then inherit the promises, through the principle of legitimacy. And there are many of the sons and daughters of Abraham among us at the present time; these will be baptized for their dead brethren and sisters, and by this means bring them unto Christ, beginning on the outside branches of the tree, and so progressing to the main stock, and from that to the root. And it shall come to pass that all Israel shall be saved. Why? Because it is their legitimate right. And they are Israel who do the works of Abraham.

Thus it is, then, with Abraham. The old man feels perfectly easy about the matter; and if he does see many of his descendants existing as a cursed race on account of their transgressions, many of them enjoying no higher avocation than crying "old clothes," still the time of their redemption will come, and by means of the eternal Gospel and priesthood, they with us will be made perfect, and we with them. While the faithful are operating in heaven to bring this about, the saints are operating on earth; and by faith and works we will accomplish all things, we will redeem the dead and the living, and all shall come forth, and Abraham will stand at the head of his seed as their ruler. This is his legitimate position.

We will now notice those men who are contending for it without any authority, and make a contrast between the two. We see them gathering their forces, and using their influence to destroy the poor among men. How long will the kings and rulers of the earth do this? Until they are dead and damned. And what then? They will be cast down into a pit. Isaiah saw them there, along with many other scoundrels, murderers and scamps. After many days they will be visited, but they have got to lie in prison a long time for their transgressions. The one is legitimacy, and the other is illegitimacy; the one is the order of God, and the other is the order of the devil.

Such is the position of things in relation to the world, to legitimacy and illegitimacy, in regard to things that are right, and things that are wrong. Jesus Christ created all things, and for him were they made, whether it be principalities, powers, thrones or dominions. Now the question is, is he going to be dispossessed of his right, because scoundrels exist in the world, and stand in power and dominion; because his subjects have rebelled against him from time to time, and usurpers have taken his place, and the dominion is given to another? Verily, no. But the time will come when the kingdom and the greatness of the kingdom under the whole heaven will be given to the saints of the Most High, and they will possess it for ever and ever.

We will now notice some of the acts of God, and some of the acts of those who have been under the dominion of Satan, those who have had dominion over the world—the proud and haughty usurpers, and the shedders of innocent blood. These are they that have lived in the world, and possessed all the good things of it. And what has been the situation of the saints in every age? All those who dared acknowledge that God lived, that this kingdom belonged to Him, that it was His right, and that He would without doubt possess it, have been trodden under foot, persecuted, cast out, hated, killed; " they wandered about in sheep-skins and goat skins; being destitute, afflicted and tormented." As one of old says, in speaking of the Jews, which of the prophets have not your fathers killed, who testified before the coming of the Just One?

This was the case in ancient days, and has been carried on in modern times. I have, with my own eyes, seen holy prophets expire, who were killed by the hands of a murderous gang of blood-thirsty assassins, because they bore the same testimony that the holy prophets did in days of old. How many more of their brethren who dared to acknowledge the truth, have fallen beneath the same influences—have been shot, whipped, imprisoned, and put to death in a variety of ways, while hundreds of others, driven from their homes in the winters, have found their last bed; they were worn out with suffering and fatigue, the weary wheels of life stood still; they were obliged to forsake the world, in which they could no longer remain, because of the persecution heaped upon them by the enemies of the truth.

The reason of all this vile outrage upon innocent men, women and children, is because there is no legitimate rule upon the earth. God's laws and government are not known, and His servants are despised and cast out.

Legitimacy and right, whether in heaven or on earth, cannot mix with anything that is not true, just and equitable; and truth is free from oppression and injustice, as is the bosom of Jehovah. Nothing but that will ultimately stand. What has been the position of the world generally, among themselves? You see men marshalling armies, and making war with one another to destroy each other, and take possession of their territory and wealth. One man who is in possession of wealth, power and authority, sees oppression exercised by kings; so he follows the example, as do rulers who exercise authority under their sovereign; then others in a still lower degree do the same; thus oppression treads upon the heels of oppression, and distress follows distress. You will find this to exist in a great measure through every grade of society, from the king on his throne, down to the match-maker or the chimney-sweep.

To ameliorate the condition of man, there are a great many institutions introduced into the world in the shape of Tract Societies, Bible Societies, and many more too numerous for me to name. Many of them are founded by sincere men, but commencing on the wrong foundation, they keep wrong all the time, and fail to accomplish the object desired. If any one of these different institutions were to carry out their own principles, they would not only fail in accomplishing the object they have in view, but ultimately destroy themselves.

There are Peace Societies among the rest; their object is to bring peace into the world, without the Spirit of God. They see plainly that peace is desirable, but they wish to graft it on to a rotten stock. In Europe they had a "Peace Congress," and sent their representatives to all parts of the world; and of course this "Congress of Peace" wished to regulate the world, make an end of war, and bring in universal peace.

Talk about peace, when rancorous discord makes its nest in the councils and cabinets of all nations, and the hearts of their statesmen are steeped in hatred one to another! Jealousy, animosity and strife, like the influence of a deadly contagion, may be found in almost every family; brother rising up against sister, sister against brother, the father against the mother, and the mother against the father, etc. We can find discord reigning even in the "Peace Society" itself.

Jesus Christ says, "My peace I give unto you; not as the world giveth, give I unto you," etc. Wherever this peace exists, it leaves an influence that is comforting and refreshing to the souls of those who partake of it. It is like the morning dew to the thirsty plant. This peace is alone the gift of God, and it can only be received from Him through obedience to His laws. If any man wishes to introduce peace into his family or among his friends, let him cultivate it in his own bosom; for sterling peace can only be had according to the legitimate rule and authority of heaven, and obedience to its laws.

Everything is disordered, and in confusion in the world. The reason is, because no legitimate authority has been known or acknowledged on the earth. Others have been trying to build up and establish what they supposed to be the kingdom of God. The socialists of France call themselves religious people, and they also expect to bring about a reign of glory through a species of Robespierreism. I was told by a man well acquainted with matters of fact in relation to these things, that if they gained the ascendancy in France, their first object would be to erect a statue to Robespierre. They were going to cut off thousands of people, to accomplish their designs; and had not Napoleon taken active measures to head them, bands of men were ready on a moment's warning to cut off the heads of thousands, and among these, I was informed, fifty thousand priests were doomed.

These are some of the principles and ideas that exist in the world, among the various nations and institutions of men, which are framed according to illegitimate principles. A change of government changes not the condition of the people, for all are wrong, and acting without God.

Our ideas are, that the time has come to favor God's people; a time about which prophets spoke in pathetic strains, and poets sung. These men of God looked through the dark vista of future ages, and being wrapped in prophetic vision, beheld the latter day glory—the time of the dispensation of the fullness of times, spoken of by all the holy prophets since the world began; for they all looked forward with joyful anticipations to the things which have commenced with us; they all had their eye upon the time when legitimacy would obtain its proper place upon the earth, in the shape of the kingdom of God established in the world, when all false rule and dominion would be put down, and the kingdoms of this world would become subject to God and His Christ. These are the ideas that they had, and these are the things we are seeking to carry out.

If we look at what illegitimacy has done in former times, we shall see the absolute necessity of the restitution spoken of by the prophets, for it has filled the earth with evil, it has caused the world to groan in bondage, laid millions in the cold embrace of death, and caused disease to spread its pestiferous breath among the nations, leaving ruin, misery and desolation in its path, and made this fair earth a howling wilderness. And nothing but the wisdom and intelligence of God can change it. The kingdom of God will establish truth and correct principles—the principles of truth, equity and justice; in short, the principles that emanate from God, principles that are calculated to elevate man in time and through all eternity. How shall this be? It will be by a legitimate rule, authority and dominion.

Who have we for our ruling power? Where and how did he obtain his authority? Or how did any of this Church and kingdom obtain it? It was first obtained by a revelation from the Lord of the universe, by the opening of the heavens, by the voice of God, and by the ministering of holy angels. It is by the voice of God and the voice of the people, that our present President obtained his authority. Many people in the world are talking about mis-rule and mis-government. If there is any form of government under the heavens where we can have legitimate rule and authority, it is among the Saints. In the first place, we have a man appointed by God, and, in the second place, by the people. This man is chosen by yourselves, and every person raises his hand to sanction the choice. Here is our President, Brigham Young, whom we made choice of yesterday, who is he? He is the legitimate ruler among this people. Can anybody dispossess him? They cannot, because it is his legitimate right, and he reigns in the hearts of the people. He obtains his authority first from God, and secondly from the people; and if a man possesses five grains of common sense, when he has a privilege of voting for or against a man, he will not vote for a man that oppresses the people; he will vote according to the dictates of his conscience, for this is the right and duty of this people in the choice of their President, and other leading officers of the kingdom of God. While this is being done here, it is being done in every part of the world, wherever the Church of Jesus Christ of latter-day saints has a footing. Is there a monarch, potentate or power under the heavens that undergoes a scrutiny as fine as this? No, there is not; and yet this is done twice a year, before all the saints in the world. Here are legitimacy and rule. You place the power in their hands to govern, dictate, regulate and put in order, the affairs of the kingdom of God. This is *Vox Dei vox populi.* God appoints; the people sustain. You do this by your own act; very well, then, it is legitimate, and must stand, and every man is bound to abide it if it takes the hair off his head. I know that there are things sometimes that are hard, tough and pinching; but if a man is a man of God, he has his eyes upon eternal things, and is aiming to accomplish the purposes of God, and all will be well with him in the end.

What advantage is there, then, between this government and others? Why we have peace, and as eternal beings we have knowledge of eternal things. While listening to the remarks made on this stand, what have we not heard—what have we not known? The curtains of heaven have been withdrawn, and we have gazed as by vision upon eternal realities. While, in the professing world, doubt and uncertainty throw their dark mantle over every mind.

Let us now notice our political position in the world. What are we going to do? We are going to possess the earth. Why? Because it belongs to Jesus Christ, and he belongs to us, and we to him; we are all one, and will take the kingdom and possess it under the whole heavens, and reign over it for ever and ever. Now, ye kings and emperors, help yourselves, if you can. This is the truth, and it may as well be told at this time as at any other.

> "There's a good time coming, Saints,
> A good time coming,
> There's a good time coming, Saints,
> Wait a little longer."

Having said so much on this point, we will return to the principle of legitimacy. God is our legitimate Father, and we are His children, and have a claim upon Him, and He has a claim upon us. We have come into this world to accomplish a certain purpose, and we have come in the dispensation of the fullness of times, when God decreed to gather all things together into one, whether they be things in heaven or on earth; and everything that has been in existence in any age of the world, or that is, or will be, which is calculated to benefit and exalt men, we shall have; consequently, it is for us to look after anything and everything that ever has been true, or that ever has been developed in any period of the history of man, for it all belongs to us, and has got to be restored, for restitution means bringing back that which was lost. If the antedeluvians enjoyed anything that was good, true and eternal, which is not yet made known to us, it has to be restored; or if anything existed among the ancient patriarchs and prophets, that has been lost, it has to be restored. If there are any people of God upon any detached part of this world, they with it have got to be restored. God's word will also be gathered into one, and His people and the Jews will hear the words of the Nephites, and the ten tribes must hear the words of the Jews and Nephites, and God's people be gathered and be one. All things will be gathered in one, and Zion be redeemed, the glory of God be revealed, and all flesh see it together. God's dominion will be established on the earth, the law go forth from Zion, and the word of the Lord from Jerusalem, and the kingdoms of this world will become subject to God and His Christ.

As eternal beings, then, we existed with our Father in the eternal worlds. We came on to this earth, and obtained tabernacles, that through taking possession of them, and passing through a scene of trial, and tribulation, and suffering, we might be exalted to more glory, dignity and power, than would have been possible for us to obtain had we not been placed in our present position. If any of you do not believe this, let me refer you to a passage of Scripture or two. How was man created at first? We are told that God made man a little lower than the angels; then says Paul, "Know ye not that we shall judge angels." What through? It is through the atonement of Jesus Christ, through the taking of our bodies, the powers of the holy priesthood, and the resurrection of Jesus Christ, that we shall obtain a higher exaltation than it would have been possible for us to enjoy, if we had not fallen. To do right in our present state, then, we must carry out the principle of legitimacy according to a correct rule, and, if we profess to be subjects of the kingdom of God, we must be subject to the dominion, rule, legitimacy, and authority of God. No person can escape from this unless he apostatizes, and goes to the devil, like a fool. He must be a fool who would barter away eternal life, thrones, principalities and powers in the eternal world, for the paltry trash which exists in the shape of wealth and worldly honor; to let go his chance of heaven and of God, of being a king and priest unto Him, of living and reigning for ever, and standing among the chiefs of Israel. I cannot help calling such men fools, for they are damned now in making such choice, and will be hereafter.

I will say a little more on legitimacy and right to rule. What would be the position of a man who would take a course to rob his neighbor, or take advantage of him in the case of his legitimacy, which you have heard of this morning? Such a man must be a greater fool than the other. For instance, a good man dies, who has served God in righteousness all his days; the weary wheels of life stand still, and he goes to the world of spirits. He believed in the principles of justice, equity, righteousness, and truth, and that his rights would be held sacred to him by his brethren after he was gone. But some professed man of God comes to his widow, and wants to steal her away from him; he would rob the dead with impunity, under the ostensible garb of justice to her and her dead husband; he will tell her he is doing it out of pure love to them both, and he is going to exalt them in the kingdom of God. We read of the kingdom of God suffering violence; if violence is ever attempted, it is in a case of this kind. It is bad enough to steal from a man his earthly property; his oxen, his cow, his horse, his harness, his wagon wheels, and other paraphernalia; but what think you of a man that would rob the dead of a treasure which he holds the most dear, and prized as the most precious thing he possessed on earth—his affectionate wife! Such a person will assuredly miss his figure.

You will find in the ancient laws of Israel, there were proper rules in relation to these matters; one was, that if a man died without a child, his brother or the nearest relation of the husband should take the widow, and raise up seed to the husband, that his name might be continued in Israel, and not be blotted out. Where did these laws come from? We are told they came from God. But instead of doing this, suppose he should try to steal this woman away, and rob his brother—how would he get along, I wonder, with such a case against him, at the bar of justice? The laws and ordinances that exist in the eternal world have their pattern in the things which are revealed to the children of men on earth. The priesthood as it exists on the earth is a pattern of things in heaven. As I said in a former part of this discourse, priesthood is legitimate rule, whether on earth or in heaven. When we have the true priesthood on earth, we take it with us into the heavens; it changes not, but continues the same in the eternal world.

There is another feature of that ancient law which I will mention. It was considered an act of injustice for the nearest relation not to take the wife of the deceased; if he refused to do it, he was obliged to go before the elders of "Israel, and his brother's wife shall loose his shoe from off his foot, and spit in his face, and shall answer and say, So shall it be done unto the man that will not build up his brother's house; and his name shall be called in Israel, The house of him who hath his shoe loosed." If the restitution of all things is to be brought to pass, there must be a restitution of these things; everything will be put right, and in its proper place.

There is another thing which is most grievous, afflicting and distressing so contemplate. When a man takes to himself a woman that properly belongs to another, and defiles her, it interferes with the fountain of life, and corrupts the very source of existence. There is an offspring comes forth as the fruit of that union, and that offspring is an eternal being—how can it be looked upon? To reflect upon it, wounds the finest feelings of human nature in time, and will in eternity. For who can gaze upon the degradation of their wife, and the corruption of their seed, without peculiar sensations? How much more is this feeling enhanced when the wronged man considers that he has been robbed by one who professed to be his friend? This thing is not to be trifled with, but is of the greatest importance; hence the necessity of the sealing powers, that all things may be pure, chastity maintained, and lasciviousness be rooted out from among the saints. Why so? That we may have a holy offspring, that shall be great, and clothed with the mighty power of God, to rule in His kingdom, and accomplish the work we propose they shall fulfill; and that when we go to sleep, we may sleep in peace, knowing that justice will be administered in righteousness. We shall know that we have a claim upon our own in the first resurrection; we shall know that our wives and our children will be there to join us, justice will be administered, and we shall have a claim upon them in the eternal world, and that no unprincipled scoundrel will be permitted to set his foot on another, or rob him of his just claims. Why is a woman sealed to a man for time and all eternity? Because there is legitimate power on earth to do it. This power will bind on earth and in heaven; it can loose on earth, and it is loosed in heaven; it can seal on earth, and it is sealed in heaven. There is a legitimate, authorized agent of God upon earth; this sealing power is regulated by him; hence what is done by that, is done right, and is recorded. When the books are opened, every one will find his proper mate, and have those that belong to him, and every one will be deprived of that which is surreptitiously obtained.

Let us do righteously, and you who would seek to injure another, and take advantage of one who was just and faithful to his God in his day, how would you like, when you get a few years older and drop into eternity, for somebody to come and serve you the same? You could not expect anything else, you could not die without being menaced by this supposition, and your dying pillow would be made unhappy, you would know you had done wrong, and would expect somebody to measure to you the same measure pressed down, shook together, and running over.

We have been told to preach confidence; correct principles and just dealings alone will inspire it. If a man speaks that which is not true about another, can you have confidence in him? No. If a man defrauds another, can you have confidence in him? No. But if you would, through a principle of covetousness, seek to sap the foundation of another's happiness, by trying to wrench from him those sacred rights which pertain to his interest in the eternal world, how much greater will be your condemnation? Nothing but truth, integrity, virtue, honor, and every pure principle, will stand in the great day of God Almighty. If such a person happens to get through this world, he will find barriers in the next, and probably miss a chance of obtaining a place in the first resurrection. Nothing contrary to the authority, rule, and government of heaven, will stand in time or in eternity; and if any man wants to be blessed and honored, and to obtain a high place in the eternal world, let him pursue a course of honor, righteousness, and virtue before his God; and if he wants to find himself amongst usurpers, defrauders, oppressors, and those in possession of illegitimate claims, let him take an opposite course. If time would permit, much more might be said about social, family and individual legitimate rights; but as time hastens, I forbear for the present.

Well, brethren and sisters, may God bless you. Amen.

THE END.